Collins

The Shanghai Maths Project

For the English National Curriculum

Series Editor: Professor Lianghuo Fan
UK Curriculum Consultant: Jo-Anne Lees

Practice Book

William Collins' dream of knowledge for all began with the publication of his first book in 1819.
A self-educated mill worker, he not only enriched millions of lives, but also founded a flourishing publishing house. Today, staying true to this spirit, Collins books are packed with inspiration, innovation and practical expertise. They place you at the centre of a world of possibility and give you exactly what you need to explore it.

Collins. Freedom to teach.

Published by Collins
An imprint of HarperCollins*Publishers*
The News Building
1 London Bridge Street
London SE1 9GF

HarperCollins*Publishers*
Macken House, 39/40 Mayor Street Upper,
Dublin 1, D01 C9W8, Ireland

Browse the complete Collins catalogue at
www.collins.co.uk

10 9 8 7 6 5 4 3

ISBN 978-0-00-814471-5

British Library Cataloguing-in-Publication Data
A catalogue record for this publication is available from the British Library.

Series Editor: Professor Lianghuo Fan
UK Curriculum Consultant: Jo-Anne Lees
Publisher: Katie Sergeant
Publishing Manager: Fiona McGlade
Project Manager: Mike Appleton
Cover designer: East China Normal University Press and Steve Evans
Cover illustrator: Steve Evans
Typesetter: East China Normal University Press
Production controller: Katharine Willard
Printed and bound in the UK by Ashford Colour Press Ltd

This book is produced from independently certified FSC™ paper
to ensure responsible forest management.

For more information visit: www.harpercollins.co.uk/green

Contents

Chapter 9   Circles and properties of circles

End of year test / 218

Answers / 224

Chapter 1　Introduction to linear inequalities

1.1　Inequalities and their properties (1)

Learning objective

Understand and apply the concepts of inequalities and their properties.

A. Multiple choice questions

1　Ben is 18 years old and Alvin is 11 years old. Read these statements. The correct one is ( ).

A. Those who are older than Alvin are definitely older than Ben.

B. Those who are younger than Ben are definitely younger than Alvin.

C. Those who are older than Ben may be younger than Alvin.

D. Those who are younger than Alvin are definitely not older than Ben.

2　Let a be a rational number greater than 1. The corresponding points of a, $\dfrac{a+2}{3}$ and $\dfrac{2a+1}{3}$ on a number line are O, P and Q, respectively. Then the order of the three points on the number line, from left to right, is ( ).

A. Q, P and O　　　B. P, Q and O　　　C. O, P and Q　　　D. Q, O and P

B. Fill in the blanks

3　Inequality property 1: When _____ is added to or subtracted from both sides of an inequality, the direction of the inequality sign remains _____.

4　Write an inequality to express "the sum of a and negative b is greater than 0": _____.

5　Write an inequality to express "the difference between 8 and 2 times y is a non-negative number": _____.

6 Write an inequality to express "the difference between half of x and 3 is less than or equal to -5": _____ .

7 Write an inequality to express "the sum of a squared and b squared is not less than 4 times the product of a and b": _____ .

8 Complete each statement with an inequality sign.

(a) When $x < y$, $x + 2$ _____ $y + 2$ and $y - x$ _____ 0.

(b) When $x > 0$ and y _____ 0, $\dfrac{x}{y} < 0$.

(c) When $x < 0$ and y _____ 0, $xy \geqslant 0$.

(d) When $4x - 3 > 2y - 3$, $4x$ _____ $2y$.

(e) When $3x - 1 < 3$, $3x$ _____ 4.

C. Questions that require solutions

9 Given that $a < b$, compare the values of $a - b$ and $b - a$.

10 Compare the values of $\dfrac{a}{9} - 8$ and $\dfrac{a}{9} - 9$.

11 Compare the values of $\dfrac{x^2 - y^2 + 1}{2}$ and $\dfrac{x^2 - 2y^2 + 1}{3}$.

12 Compare the values of $\dfrac{2}{3}a + 5$ and $3 - \dfrac{1}{3}a$.

1.2   Inequalities and their properties (2)

Learning objective

Understand and apply the concepts of inequalities and their properties.

A. Multiple choice questions

1 Read these statements. The correct one is (   ).

A. Given that $a^2 > 0$, then $a > 0$.　　　B. Given that $a^2 > a$, then $a > 0$.

C. Given that $a < 1$, then $a^2 < a$.　　　D. Given that $a < 0$, then $a^2 > a$.

2 Read these statements. (   ) of them is/are correct.

① Given that $-4x > 20$, then $x > -5$.　　② Given that $a < b$, then $ac < bc$.

③ Given that $a > b$, then $ac^2 > bc^2$.　　④ Given that $ac^2 > bc^2$, then $a > b$.

A. one　　　　　B. two　　　　　C. three　　　　　D. none

3 Given that $x > y$, the condition satisfying $ax \leqslant ay$ is (   ).

A. $a > 0$　　　B. $a < 0$　　　C. $a \geqslant 0$　　　D. $a \leqslant 0$

B. Fill in the blanks

4 **Inequality property 2:** When both sides of an inequality are multiplied or divided by
_____ , the direction of the inequality sign remains unchanged.

5 **Inequality property 3:** When both sides of an inequality are multiplied or divided by
_____ , the direction of the inequality sign is changed.

6 The diagram shows the positions of rational numbers on a number line. Fill in the
blanks with inequality signs.

(a) $a + c$ _____ $b + c$　　　　(b) ab _____ 0

(c) $\dfrac{a}{b}$ _____ 0　　　　(d) $\dfrac{c}{b}$ _____ 1

Diagram for question 6

7 Fill in the blanks with inequality signs.

(a) Given that $-3x > -3y$, then $-12x$ _____ $-12y$

(b) Given that $a < 0$ and $b < c$, then ab _____ ac

(c) Given that $x - 2y > x$, then y _____ 0

(d) Given that $x < y$, then $-\dfrac{2}{3}x + 1$ _____ $-\dfrac{2}{3}y + 1$

(e) Given that $a > 0$, $b < 0$ and $c < 0$, then $(a - b)c$ _____ 0

C. Questions that require solutions

8 Write a valid inequality based on each given condition.

(a) Given that $a < 2b$, subtract c from both sides of the inequality.

(b) Given that $a < 2b$, multiply both sides of the inequality by c^2.

(c) Given that $a < 2b$, divide both sides of the inequality by $c - 1$ ($c \neq 1$).

9* Given that $a < 0$, $b > 0$ and $a + b < 0$, use " $<$ " to put $-a$, $-b$, $-\sqrt{a^2}$ and $a - b$ in order.

10* Given that $m < 0$ and $-1 < n < 0$, put m, mn and mn^2 in order from the least to the greatest.

---

* Challenging questions; pay careful attention.

1.3 Solving linear inequalities in one variable (1)

Learning objective

Solve linear inequalities in one variable; represent the solution set on a number line.

A. Multiple choice questions

1 Of these statements, ( ) is/are correct.

① 3 is a solution to the inequality $x - 1 > 1$.

② $x > 5$ is the solution set of the inequality $x + 4 > 8$.

③ -1 is a solution to the inequality $x + 1 \geqslant 0$.

④ There are infinitely many positive integer solutions to the inequality $x < 11$.

A. one   B. two   C. three   D. four

2 When x takes a value not greater than 2.5, the value of $2x - 5$ is ( ).

A. a positive number

B. less than or equal to 0

C. a negative number

D. a non-negative number

3 Of these inequalities, the one that does not hold true is ( ).

A. $-7.8 < -5.5$ B. $21 > 13$ C. $1 + a^2 > 0$ D. $m > -m$

4 Of these representations of the solution set to inequality $3x - 1 < 5$ on the number line, the correct one is ( ).

A.
```
     ┌───────●
  ┼──┼──┼──┼──┼──┼──▶
 -2 -1  0  1  2  3
```

B.
```
            ┌─────────┐
  ──────────────────○──┼──▶
  -2 -1  0  1  2  3
```

C.
```
        ┌──────────
  ┼──┼──┼──┼──┼──┼──▶
  0  1  2  3  4  5
```

D.
```
        ┌──────────
  ┼──┼──○──┼──┼──┼──▶
  0  1  2  3  4  5
```

Diagram for question 4

B. Fill in the blanks

5 In an inequality with variables, _____ _____ are called the solutions to the inequality.

6 The collection of all solutions to an inequality is called the _____ of the inequality.

7 The process of finding the _____ is called solving the inequality.

8 The equation $5x = 10$ has _____ solution(s). The inequality $5x < 10$ has _____ _____ solution(s).

9 If the solution set of the inequality $(a - 3)x > a - 3$, in which x is a variable, is $x < 1$, then the set of values that a can take is _____.

C. Questions that require solutions

10 Find the solution set for each inequality.

(a) $x + 3 \geqslant -2$

(b) $-x + 3 > 2$

(c) $-3 - x > 2$

(d) $3x > -2$

(e) $-3x \leqslant -2$

(f) $-3x \geqslant 2$

11 Find the solution set for each inequality and represent it on a number line.

(a) $5x - 1 < 3\dfrac{1}{2}$

(b) $0.7x + 2.4 \geqslant 4\dfrac{1}{2}$

(c) $-\dfrac{6}{7}x - 1.2 < 3\dfrac{3}{5}$

(d) $3.25 - \dfrac{3}{4}x \geqslant 5\dfrac{1}{2}$

12 What positive integer value should a take, in order that the solution for the equation $5x = a - 11$ is a negative integer?

13 To complete an order, a toymaker planned to make 176 toys in July. For the first 10 days, he averaged 4 toys per day. After an improvement in work efficiency, he completed the order 3 days ahead of schedule. Assume that he made x toys each day, on average, after the first 10 days.

(a) Write a mathematical expression that x satisfies.

(b) Find the minimum value of x.

1.4 Solving linear inequalities in one variable (2)

Learning objective

Solve linear inequalities in one variable; use set notation to represent the solution set.

A. Multiple choice questions

1. In the following inequalities, the linear inequality in one variable is (   ).

   A. $3x - 5y < 1$ B. $x^2 - 4x > 0$ C. $\dfrac{3x - 1}{4} - 1 \geqslant 0$ D. $3 - \dfrac{1 - x}{x} \leqslant 0$

2. Given that the value of $2x + 5$ is a non-positive number, then the set of values that x can take is (   ).

   A. $x < -\dfrac{5}{2}$ B. $x \leqslant -\dfrac{5}{2}$ C. $x < \dfrac{5}{2}$ D. $x \geqslant -\dfrac{5}{2}$

3. There are (   ) non-positive integer solutions to the inequality $3x + 6 > -3$.

   A. 1   B. 2   C. 3   D. 4

B. Fill in the blanks

4. An inequality that has one _____ , and the _____ of the terms in the variable is 1 is called a linear inequality in one variable.

5. When $k =$ _____ , the inequality $(k + 3)x^{k^2 - 8} > 0$ is a linear inequality in one variable x .

6. The non-negative integer solutions to the inequality $3x - 5 > 5x - 13$ are _____ .

7. If the sum of three consecutive positive integers is less than 14, then these three consecutive positive integers are _____ .

8. If the positive integer solutions to the inequality $2x - a \leqslant 0$ are $x = 1$, $x = 2$ and $x = 3$, then the set of values that a can take is _____ .

C. Questions that require solutions

9 Solve these inequalities.

(a) $5(x-2) > 4(2x-1)$

(b) $2(1-x) + 3(-1-x) \geqslant 6 - (3x+2)$

(c) $5 - 4[3(x-2)-1] < 57$

10 Given that $3 - (a-4) > 3a - 1$, simplify $\sqrt{(a-4)^2} - \sqrt{(4-2a)^2}$.

11 There are 20 questions in the first round of a general knowledge competition. According to the rules, 10 points are awarded for each correct answer, 5 marks are deducted for each incorrect answer or unanswered question, and those with a total score not less than 80 marks will pass to the next round of the competition. Sam passed the first round. At least how many questions did he answer correctly?

12 The smallest integer solution to the inequality $4(x-3) + 5 < 6(x-2) + 1$ is the solution to the equation $4x - ax = 3$. Find the value of a.

1.5 Solving linear inequalities in one variable (3)

Learning objective

Solve challenging linear inequalities in one variable.

A. Multiple choice questions

1 The integer solution to the inequality $x - \dfrac{3}{2} \geqslant \dfrac{x}{2} - \dfrac{1}{3}$ is (　　).

    A. an integer greater than 2   B. an integer not less than 2

    C. 2   D. $x \geqslant 3$

2 If the solution set of the inequality $\dfrac{ax - 5}{2} - \dfrac{2 - ax}{4} > 0$ is $x > 1$, then the value of a is

    (　　).

    A. 3   B. 4   C. −4   D. none of the above

B. Fill in the blanks

3 The solution set of the inequality $\dfrac{3x - 2}{2} > \dfrac{x - 1}{3}$ is _____.

4 When the value of $\dfrac{2(2x - 3)}{3}$ is not a positive number, the set of values that x can take is _____.

5 Given that the value of $\dfrac{3}{2}y - \dfrac{2}{3}(y + 1)$ is greater than 1, the set of values that y can take is _____.

6 When the value of $\dfrac{1 - 2m}{5}$ is not less than the value of $3m + 2$, the set of values that m can take is _____.

7 Given that the solution to the equation $(1 - m)x = 1 - 2x$ in one variable x is a negative number, then the set of values that m can take is _____.

C. Questions that require solutions

8 Solve these inequalities.

(a) $\dfrac{5}{2} - \dfrac{3}{4}(x + 5) < \dfrac{5}{8}(x - 2) - 2$ (b) $x + 1 - \dfrac{x - 1}{2} > \dfrac{4x}{3} - 1$

9 Find the smallest integer that satisfies the inequality $\dfrac{2x - 5}{3} < \dfrac{6x - 1}{4}$.

10 Given that the solution to the equation $\dfrac{2}{3}x - 1 = 6m + 5(x - m)$ in one variable x is a non-negative number, find the set of values that m can take.

11 Given that the solution to the equation $5x - 2k = -x + 4$ in one variable x is greater than 1, find the set of values that k can take.

12 The solution sets of the inequalities $\dfrac{2x - a}{3} > \dfrac{a}{2} - 1$ and $\dfrac{x}{a} < 2$ are exactly the same. Find the value of a.

1.6 Simultaneous linear inequalities in one variable (1)

Learning objective

Recognise and solve simultaneous linear inequalities in one variable.

A. Multiple choice questions

1 The solution set of simultaneous inequalities
$\begin{cases} 2x > -3 \\ x < 4 \end{cases}$ is ( ) *.

A. $x < -\dfrac{3}{2}$ B. $x > 4$

C. $-\dfrac{3}{2} < x < 4$ D. impossible to find

> A set of simultaneous inequalities is also known as a system of inequalities.

2 Given that the solution set of simultaneous inequalities $\begin{cases} x > 2 \\ x > a \end{cases}$ in one variable x is $x > 2$, then the set of values that a can take is ( ).

A. $a > 2$ B. $a < 2$ C. $a \geqslant 2$ D. $a \leqslant 2$

3 The sum of a number and 2 is greater than 1, and the difference between twice the number and 3 is not greater than 5. A system of inequalities that satisfy the conditions is ( ).

A. $\begin{cases} x + 2 > 1 \\ 2x - 3 < 5 \end{cases}$ B. $\begin{cases} x + 2 > 1 \\ 2x - 3 > 5 \end{cases}$ C. $\begin{cases} x + 2 \geqslant 1 \\ 2x - 3 \leqslant 5 \end{cases}$ D. $\begin{cases} x + 2 > 1 \\ 2x - 3 \leqslant 5 \end{cases}$

B. Fill in the blanks

4 A system of inequalities that consists of several _____ is called a system of linear inequalities in one variable.

5 The area of a piece of green land is 900 square units and the length can be from 75 units to 100 units, then the range of the width x is _____.

---

* When solving simultaneous inequalities, you may first use a number line to represent the solution to each inequality in order to find the solution set.

6 The solution set of the system of inequalities $\begin{cases} x \geqslant 4 \\ x > 6 \end{cases}$ is _____.

7 The solution set of the system of inequalities $\begin{cases} x < 3 \\ x \geqslant -2 \end{cases}$ is _____.

8 Given that the simultaneous inequalities $\begin{cases} x > a \\ x \leqslant 3 \end{cases}$ has a solution, then the set of values of a is _____.

C. Questions that require solutions

9 Solve these simultaneous inequalities.

(a) $\begin{cases} 3x < 4x \\ 5x - 4 \geqslant 2 + 7x \end{cases}$

(b) $\begin{cases} 2x + 1 < 3 \\ 3(x - 2) - 1 \geqslant 2(x - 4) \end{cases}$

10 Find the non-negative integer solutions to these simultaneous inequalities.

$$\begin{cases} 2(x + 2) + 1 > -3 \\ -1 + 2x < 8 - \dfrac{x}{4} \end{cases}$$

11 Given that the solution set to the simultaneous inequalities $\begin{cases} x + 5 > 2a \\ x + a < 3b \end{cases}$ in the variable x is $-9 < x < 10$, find the values of a and b.

12 Given that the solution set to the system of inequalities $\begin{cases} 2x + 2 > 4(x - 2) \\ x + a < 1 \end{cases}$ in the variable x is $x < 5$, find the set of values that a can take.

1.7 Simultaneous linear inequalities in one variable (2)

Learning objective

Solve simultaneous linear inequalities in one variable.

A. Multiple choice questions

1. Of these simultaneous inequalities, the one that has no solution is (   ).

   A. $\begin{cases} x + 2 > 0 \\ x + 1 < 0 \end{cases}$
   B. $\begin{cases} x + 2 > 0 \\ x + 1 > 0 \end{cases}$
   C. $\begin{cases} x + 2 < 0 \\ x + 1 > 0 \end{cases}$
   D. $\begin{cases} x + 2 < 0 \\ x + 1 < 0 \end{cases}$

2. The solution set to the simultaneous inequalities $\begin{cases} x > -2 \\ x < 3 \\ x > 1 \end{cases}$ is (   ).

   A. $x > -2$
   B. $x < 3$
   C. $-2 < x < 3$
   D. $1 < x < 3$

3. Given that the solution set to the simultaneous inequalities $\begin{cases} x + a > 0, \\ x - b < 0 \end{cases}$ is $-a < x < b$,

   then the solution set to the simultaneous inequalities $\begin{cases} x - a < 0 \\ x + b > 0 \end{cases}$ is (   ).

   A. $-b < x < a$
   B. $a < x < -b$
   C. $x < a$
   D. $x > -b$

B. Fill in the blanks

4. Given that $a > b > c$, then the solution set to the simultaneous inequalities $\begin{cases} x < a \\ x > b \\ x > c \end{cases}$ in the

   variable x is _____ .

5. Given that $a > b > c$, then the solution set to the simultaneous inequalities $\begin{cases} x < a \\ x < b \\ x > c \end{cases}$ in the

   variable x is _____ .

6 In all the integer solutions to the inequality $-7 < 2x + 5 < 1$, the value of x that is greater than -4 is _____.

7 The positive integer solution to the simultaneous inequalities $\begin{cases} 2x - 1 < x + 1 \\ x + 8 > 4x - 1 \end{cases}$ is _____.

8 In the values of x that satisfy $\dfrac{2 + x}{2} \geqslant \dfrac{2x - 1}{3}$, the sum of the integers with values not less than -9 but not greater than 9 is _____.

C. Questions that require solutions

9 Find the integer solutions to this system of inequalities: $-3 \leqslant \dfrac{-3x + 1}{2} \leqslant 2$

10 Solve each system of inequalities.

(a) $\begin{cases} 5x - 4 \leqslant 4x + 1 \\ \dfrac{x}{4} - 3 \leqslant \dfrac{3x + 5}{2} - \dfrac{x}{3} \end{cases}$

(b) $\begin{cases} 2x - 1 \geqslant 5 \\ \dfrac{x + 10}{2} < 7 \\ 1 - \dfrac{x + 1}{2} \geqslant 3 - 2(x - 2) \end{cases}$

11 A group of children are to share some toys. If each child gets 4 toys, then there will be 27 toys left over. If each child gets 5 toys, then there are not enough for the last person. What is the lowest possible number of children in the group?

12 The lengths of the three sides of a triangle are $4a + 5$, $2a - 1$ and $20 - a$ (a is an integer). Find the perimeter of the triangle.

Unit test 1

 A. Multiple choice questions

1 Read these statements. The correct one is (   ).

    A. If $x > -1$, then $3x > 0$.
                                 B. If $x > -1$, then $3x < 0$.

    C. If $x > -1$, then $3x > -3$.
                         D. If $x > -1$, then $3x < -3$.

2 Given that the solution set to the inequality $(a - 2)x < a - 2$ is $x > 1$, then the set of values that a can take is (   ).

    A. $a < 0$ B. $a > 0$ C. $a < 2$ D. $a > 2$

3 Given that there is a solution to the simultaneous inequalities $\begin{cases} -1 < x \leqslant 3 \\ x > m \end{cases}$, then the set of values that m can take is (   ).

    A. $m \leqslant 3$ B. $m < 3$ C. $m < -1$ D. $-1 < m < 3$

4 Given that $a > b$ then, of these inequalities, the one that is true is (   ).

    A. $a^2 > b^2$ B. $a - 3 > b - 3$ C. $-3a > -3b$ D. $3 - a > 3 - b$

5 In the process of solving the inequality $\dfrac{x - 1}{3} - \dfrac{2x + 3}{2} < \dfrac{x}{6} - 1$, (   ) is incorrect.

    Step 1: $2(x - 1) - 3(2x + 3) < x - 1$ Step 2: $2x - 2 - 6x - 9 < x - 1$

    Step 3: $-5x < 10$ Step 4: $x > -2$

    A. Step 1   B. Step 2   C. Step 3   D. Step 4

 B. Fill in the blanks

6 Given that $a < b$, then $-3a - 2$ _____ $-3b - 2$. (Write " $>$ " or " $<$ ".)

7 Given that the solution set of $(1 - 3a)x \geqslant 1$ is $x \leqslant \dfrac{1}{1 - 3a}$, then the set of values that a can take is _____.

8 The solution set to the system of inequalities $\begin{cases} 2x + 3 > 5 \\ -3x + 2 < -4 \end{cases}$ is _____.

9 Given that the values of both algebraic expressions $5x + 6$ and $3x + 1$ are greater than 0, then the set of values that x can take is _____.

10 Given that there are only three positive integer solutions to the inequality $3x - m \leqslant 0$ in x, then the set of values of m is _____.

11 The solution set of the inequalities $\begin{cases} 2x > -1 \\ 2(x - 1) \leqslant 5 \end{cases}$ is _____.

12 Given that $3(4x + 1) + 3 < 6x$, then $5x - 2 <$ _____.

C. Questions that require solutions

13 Solve the equality: $\dfrac{x - 1}{3} - \dfrac{2x + 4}{4} > -3.$

14 Solve the inequality: $\dfrac{x - 2}{5} - \dfrac{x + 3}{10} > \dfrac{2x - 5}{3} - 3.$

15 Find the integer solution to the system of inequalities $\begin{cases} 2(x + 2) < x + 5 \\ 3(x - 2) + 8 \geqslant 2x \end{cases}.$

16 The parking charges at two city car parks are as follows:

- the first car park charges a flat rate of £2.50 per hour
- the second car park charges £3 for the first hour, and then £2 per hour subsequently.

(a) If a driver needs to park his car for 3 hours, which car park should he choose to be charged less? How much can he save?

(b) After how many hours of parking will the parking charge at the second car park be cheaper than at the first car park? Use an inequality to solve this question.

17 First find the solution set of the system of inequalities $-2 < \dfrac{1}{2}x - 1 < 1$, and then simplify $\sqrt{(2x + 4)^2} + 2\sqrt{(x - 4)^2}$.

18 Given that the integer solution to the simultaneous inequalities $\begin{cases} 2x > 3x - 3 \\ 3x - a > -6 \end{cases}$ in x is 2, find the set of values that a can take.

Chapter 2　Simultaneous linear equations

2.1　Linear equations in two variables

 Learning objective

Understand and apply relevant concepts about linear equations in two variables.

 A. Multiple choice questions

1 Rearranging the linear equation in two variables $3x - 2y = -4$ to make y the subject, the correct expression for y in terms of x is (　　).

A. $y = \dfrac{3x - 4}{2}$　　　　B. $y = \dfrac{3x + 4}{2}$　　　　C. $y = \dfrac{-3x + 4}{2}$　　　　D. $y = \dfrac{-3x - 4}{2}$

2 Given that $\begin{cases} x = 3 \\ y = 5 \end{cases}$ is a solution to the equation $mx - 2y = 2$, then the value of m is

(　　).

A. $\dfrac{8}{5}$　　　　　　B. $\dfrac{5}{3}$　　　　　　C. 4　　　　　　D. $-\dfrac{8}{3}$

3 The number of positive integer solution(s) to the equation $2x + y = 8$ is (　　).

A. 4　　　　　　　B. 3　　　　　　C. 2　　　　　　D. 1

 B. Fill in the blanks

4 A linear equation in two variables is _____.

5 Given a linear equation in two variables, the values of the two variables that make _____ sides of the equation _____ are solutions to the equation.

6 Given that the equation $2x^{a-1} + 3y^{3b-11} = -4$ is a linear equation in two variables x and y, then the value of $a + b$ is _____.

7 Given that $\dfrac{1}{2}x - \dfrac{3}{2}y = 1$, the value of y when $x = 1$ is _____.

8 Given that the sum of the solutions, x and y, to the linear equation in two variables $3x + 4y = 6$ is zero, then this solution is _____ .

C. Questions that require solutions

9 Change the subject of the equation $3x - 5y = 9$ as indicated.

(a) Make y the subject.   (b) Make x the subject.

10 Find all the positive integer solutions to the linear equation in two variables, $4x + y = 19$.

11 Find all the negative integer solutions to the linear equation in two variables, $3x + 4y = -25$.

12 Given the simultaneous equations $\begin{cases} x = t - 1 \\ y = 5t + 4 \end{cases}$, find an algebraic expression in x to express y.

2.2 Simultaneous linear equations in two variables and their solutions (1)

Learning objective

Recognise simultaneous linear equations in two variables and solve them by using substitution.

A. Multiple choice questions

1. Of these equation systems, (   ) is a system of linear equations in two variables.

   A. $\begin{cases} 2 : x^2 = 3 : y \\ 2x + y = 9 \end{cases}$

   B. $\begin{cases} y = 2x - 3 \\ xy = 1 \end{cases}$

   C. $\begin{cases} x + y = 5 \\ x - 2y = -4 \end{cases}$

   D. $\begin{cases} x + 2y = 5 \\ y + z = 4 \end{cases}$

   A set of simultaneous equations is also known as a system of equations.

B. Fill in the blanks

2. If there are _____ variables in a system of equations and the highest degrees of any terms containing the variables is _____ , then it is a system of linear equations in two variables.

3. In a system of linear equations in two variables, _____ is called a solution to the system of linear equations in two variables.

4. One way of solving simultaneous linear equations in two variables is to rearrange one equation to express either variable in terms of the other, and then _____ the expression obtained into the other linear equations to get a linear equation in _____. This is called the substitution method.

5. Given that $\begin{cases} x = 2 \\ y = -1 \end{cases}$ and $\begin{cases} x = -1 \\ y = 2 \end{cases}$ are the solutions to the equation $y = kx + b$, then the value of k is _____ and the value of b is _____.

6. Given that $\sqrt{(x + y - 1)^2} + \sqrt{(x - y + 3)^2} = 0$, then the value of $(x + y)^{2016}$ is _____.

C. Questions that require solutions

7 Use the substitution method to solve these simultaneous equations.

(a) $\begin{cases} y = 1 - x \\ 3x + 2y = 5 \end{cases}$

(b) $\begin{cases} 2x + y = 5 \\ 4x + 5y = 13 \end{cases}$

8 Solve these simultaneous equations.

(a) $\begin{cases} 3x - y = 2 \\ x + 7y = -3 \end{cases}$

(b) $\begin{cases} \dfrac{x}{6} + \dfrac{y}{2} = \dfrac{1}{6} \\ \dfrac{x}{6} + y = \dfrac{2}{3} \end{cases}$

9 Solve these simultaneous equations.

(a) $\begin{cases} 4x + 7\left(y + \dfrac{10}{11}\right) = 1 \\ 5x - \left(y + \dfrac{10}{11}\right) = 11 \end{cases}$

(b) $\begin{cases} 4(x - 0.123) + 7(y + 3.45) = 1 \\ 5(x - 0.123) - (y + 3.45) = 11 \end{cases}$

10 The solution to the systems of equations $\begin{cases} 2x - y = 1 \\ 3x + 3my = -21 \end{cases}$ is also a solution to the linear equation in two variables, $4x + y = 17$. Find the value of m.

11 Given that the system of equations $\begin{cases} 2x - y = 7 \\ ax + y = b \end{cases}$ and the system of equations $\begin{cases} x + by = a \\ 3x + y = 8 \end{cases}$ have the same solution, find the values of a and b.

2.3 Simultaneous linear equations in two variables and their solutions (2)

Learning objective

Solve simultaneous linear equations in two variables by using elimination and substitution.

A. Multiple choice questions

1 Given that the solution to the simultaneous linear equations in two variables $\begin{cases} 9x + 4y = 1 \\ x + 6y = -11 \end{cases}$

satisfies $2x - ky = 0$, then the value of k is (   ).

A. 4   B. –4   C. 1   D. –1

2 Consider the simultaneous equations $\begin{cases} 2x - \dfrac{1}{5}y = 7 \\ 10x - y = 35 \end{cases}$. The correct statement is (   ).

A. The solution is $\begin{cases} x = -4 \\ y = 5 \end{cases}$.

B. The solution is $\begin{cases} x = 5 \\ y = 4 \end{cases}$.

C. It has no solution.

D. It has infinitely many pairs of solutions.

B. Fill in the blanks

3 Two commonly used methods to solve simultaneous linear equations in two variables are
_____ and _____.

4 Given that $x = -1$ and $y = 1$ satisfy the system of equations $\begin{cases} 3x - 2y = a + 1 \\ x + 3y = 2b - 3a \end{cases}$, then the

value of $a + b$ is _____.

5 Given that one pair of solutions to the equation $4x - 3y = 7$ is $\begin{cases} x = a \\ y = b \end{cases}$, and b is 1 more

than 3 times a, then the value of a is _____ and the value of b is _____.

6 Given that $x + 6 = y - x - 8 = 2x + 3y - 1$, then the value of $3x - y$ is _____.

7 Given that $\sqrt{(x + y - 2)^2} + \sqrt{(2x - 3y + 6)^2} = 0$, then $x + 2y =$ _____.

 ## C. Questions that require solutions

8 Solve these simultaneous equations.

(a) $\begin{cases} 9x + 2y = -1 \\ 9x - 3y = 9 \end{cases}$

(b) $\begin{cases} 2x + 3y = 1 \\ 3x - 5y = 68 \end{cases}$

9 Solve these simultaneous equations.

(a) $\begin{cases} \dfrac{x + y}{2} + \dfrac{x - y}{4} = 4 \\ 4(x + y) - 3(x - y) = 12 \end{cases}$

(b) $\begin{cases} 30\% x + 40\% y = 5.3 \\ 60\% x - 50\% y = 0.2 \end{cases}$

10 Use a suitable method to solve these simultaneous equations.

(a) $\begin{cases} \dfrac{x - 5}{2} + \dfrac{y - 3}{4} = 5 \\ \dfrac{x - 5}{6} + \dfrac{y - 3}{4} = 3 \end{cases}$

(b) $\begin{cases} 53x + 47y = 112 \\ 47x + 53y = 88 \end{cases}$

11 Given that the sum of the pair of solutions, x and y, to the simultaneous equations $\begin{cases} 3x - 4y = 5k + 11 \\ 2x + 3y = k - 5 \end{cases}$ is zero, find the value of k and hence the solution to the equation system.

2.4 Applications of simultaneous linear equations (1)

 Learning objective

Apply simultaneous linear equations to solve real-life problems.

1. Students in Class A and Class B are participating in a tree-planting activity. The number of trees that Class A has planted is 15 more than half the number Class B has planted. Class B has planted 6 more trees than Class A. How many trees has each class planted?

2. A farmer harvested a total of 5730 kg of wheat from two plots of land last year. This year, he used higher-yield seeds and the production of wheat increased 10% in the first plot and 8% in the second plot. The total production of wheat from the two plots was 6240 kg this year. How many kilograms of wheat did each plot yield last year?

3. There are 26 workers in a workshop that produces two types of component, A and B. Each worker can produce either 15 type-A components or 10 type-B components per day. A set of two type-A components and three type-B components is required for a particular instrument for a model of car. How many workers should be assigned to producing the two types of component so that all the components produced each day will form a whole number of sets as required, without any being left over?

4. Two year groups, A and B, each have 44 students. Some students from both year groups joined extra-curricular clubs. The number of students joining the astronomy club from Group A is exactly $\frac{1}{3}$ of the number of students **not** joining the astronomy club in Group B. The number of students joining the astronomy club from Group B is exactly $\frac{1}{4}$ of the number of students **not** joining the astronomy club in Group A. How many students in each year group did not join the astronomy club?

5 100 red pens and 200 blue pens are put into small pouches. There are 2 red pens and 5 blue pens in each pouch. Each student in Class A received one pouch. The number of blue pens left over is 4 more than the number of red pens left over. Find the number of students in Class A.

6 The length of a railway tunnel is 1000 m. When a train passed through the tunnel, the time from the moment the front of the train entered the tunnel until the train was completely out of the tunnel was 1 minute. The time for which the whole train was in the tunnel was 40 seconds. Find the speed of the train and the length of the train.

7 A council has bought some trees that are to be planted at regular intervals along a road and at both ends. If one tree is planted every 3 m, 3 trees will be left over. If one tree is planted every 2.5 m, then the council will need 77 more trees. How long is the road? How many trees can be planted along the road?

8 * A vehicle travelled from place A to place B, which are 142 km apart. Assume the speed of the vehicle under the same road conditions is constant, and it travelled at 30 km per hour on flat roads, 28 km per hour on uphill roads and 35 km per hour on downhill roads. The outward trip took 4 hours and 30 minutes, and the return trip took 4 hours and 42 minutes. How long were the sections of flat road, uphill road and downhill road respectively in the outward trip?

---

* This is a more challenging question (optional). You might want to discuss it with your friends or teacher.

2.5 Applications of simultaneous linear equations (2)

 Learning objective

Apply simultaneous linear equations to solve real-life problems.

1 There are 37 people in Group A and 23 people in Group B. Today, the same number of people from each group need to be assigned for other tasks, which means that the number of people remaining in Group A is twice the number remaining in Group B. How many people are assigned to other tasks from Group A and Group B, respectively?

2 Altogether, there are 420 students in the three year groups. There are four more students in Year 6 than in Year 7. The number of students in Year 8 is $\frac{2}{3}$ of the number in Year 7. Find the number of students in each year group.

3 Two people, Alex and Bobbie, make the same components.

If Alex starts working 1 day earlier than Bobbie, and then they both work together for five days, then they will each make the same number of components.

If Alex makes 30 components on the first day and afterwards both work together for four days, then Bobbie will make 10 more components than Alex. Find the number of components each person makes per day.

4 The sum of the digits in a two-digit number is 3. If the two digits are swapped to obtain a new number, the new number is 9 less than the original number. Find the original two-digit number.

5 A square table comprises a top and four legs. Given that 50 table tops or 300 table legs can be made from 1 m^3 of timber, how many such square tables can be made with 5 m^3 of timber?

6 A factory has an order for 190 components. If Person A works for 2 days, and is then joined by Person B and they work together for 3 days, they can finish the task exactly. If Person B works for the first 3 days, then is joined by Person A and they work together for 2 days, they can also finish the task exactly. How many components can they each make every day?

7 Figure ① shows two types of small open-topped cuboid box, A and B, made from cardboard. Figure ② shows the square and rectangular faces of the boxes. How many small boxes of types A and B respectively can be made from 150 square sheets of cardboard and 300 rectangular sheets?

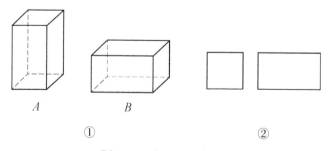

A   B

①   ②

Diagram for question 8

Unit test 2

A. Multiple choice questions

1. When the linear equation in two variables, $2x - y = 1$, is rearranged to make y the subject, the correct expression for y in terms of x is (　　).

   A. $x = \dfrac{y + 1}{2}$　　　B. $x = \dfrac{-y + 1}{2}$　　　C. $-y = 1 - 2x$　　　D. $y = 2x - 1$

2. A solution to the linear equation in two variables $2x + 3y = 14$ is (　　).

   A. $x = 4$　　　B. $y = 2$　　　C. $\begin{cases} x = 4 \\ y = 2 \end{cases}$　　　D. $\begin{cases} x = 2 \\ y = 4 \end{cases}$

3. Given simultaneous equations $\begin{cases} x - y = 2 \\ 2x + y = 7 \end{cases}$, then x is equal to (　　).

   A. 5　　　　　B. 4　　　　　C. 3　　　　　D. 2

4. Given that $\begin{cases} x = 3 \\ y = 1 \end{cases}$ is the solution to the system of equations $\begin{cases} ax - by = 10 \\ x + by = 2 \end{cases}$, then (　　).

   A. $\begin{cases} a = 3 \\ b = 1 \end{cases}$　　　B. $\begin{cases} a = -3 \\ b = -1 \end{cases}$　　　C. $\begin{cases} a = -3 \\ b = 1 \end{cases}$　　　D. $\begin{cases} a = 3 \\ b = -1 \end{cases}$

5. If $2x^{2a+b-3} + y^{a+b} = 3$ is a linear equation in two variables in x and y, then $a + 2b$ is (　　).

   A. 1　　　　　B. -1　　　　　C. 0　　　　　D. uncertain

B. Fill in the blanks

6. Given that $x = 3$ and $y = -5$ are the solutions to the equation $ax - 5y = 50$, then $a = \underline{\hspace{2cm}}$.

7. Rewrite the equation $\dfrac{x}{2} - \dfrac{y}{3} = 2$ to make y the subject of the new equation: $\underline{\hspace{2cm}}$.

8 The non-negative integer solution to the equation $3x + 2y = 10$ is _____ .

9 If $(3a + 2b - 2)^2 + (3a - 2b + 4)^2 = 0$, then the value of $a^2 + b^2$ is _____ .

10 In the linear equation in two variables $2x - 3y = 15$, if x and y add to zero, then $x =$ _____ .

11 If $2x^{m+n} - 3y^{n-2} = -7$ is a linear equation in two variables, then $m =$ _____ and $n =$ _____ .

12 There are _____ two-digit numbers with a digit sum of 6.

13 Given that the solutions to the simultaneous equations $\begin{cases} 3x + 2y = 2a \\ 4x - 3y = 6a + 2 \end{cases}$, in x and y, add to zero, then $a =$ _____ .

14 Given that the solution to the system of equations $\begin{cases} x - 3y = 12 + m \\ x + 3y = -4 - 3m \end{cases}$ is $x > 0$ and $y < 0$, and m is a positive integer, then m is _____ .

C. Questions that require solutions

15 Solve the simultaneous equations $\begin{cases} 17x - 18y = 35 \\ 7x - 8y = 15 \end{cases}$.

16 Use both the substitution method and the elimination method to solve simultaneous equations $\begin{cases} 2x + y = -1 \\ 3x + 2y = 2 \end{cases}$.

17 Given that $(3a - b + 1)^2 + (2a + 3b - 25)^2 = 0$, find the solution set for the system of inequalities $\begin{cases} 2ax - 7(x - b) > 19 \\ \dfrac{ax}{2} + (3 - b)x > 6 \end{cases}$.

18 A troop of 120 monkeys, consisting of adult monkeys and baby monkeys, shares 180 peaches. If each adult monkey gets 4 peaches and every 4 baby monkeys get 1 peach between them, then how many adult monkeys and how many baby monkeys are there?

19 Two people, A and B, set out at the same time, walking towards each other from two places that are 18 km apart. They met on the road after 2 hours. If A had started 40 minutes earlier than B, then they would have met one and half hours after B started. How fast did each of them walk, respectively, in kilometres per hour?

20 12 football teams had play-off games (one match only between every two teams). 3 points were awarded to the team winning a match, no points were awarded to the losing team, and 1 point was awarded to each team if the match was drawn. A team that had twice as many drawn matches as lost matches finally got 19 points. How many matches did the team win, draw and lose, respectively?

Chapter 3   Laws of indices and factorisation

3.1   Power of a power

Learning objective

Understand and use the meaning of the power of a power and the related law of indices.

A. Multiple choice questions

① Of these calculations, the correction one is (   ).

  A. $(a^m)^n = a^{m+n}$ B. $(a^m)^n = a^{m-n}$

  C. $(a^m)^n = a^{mn}$ D. $(a^m)^n = a^{m \div n}$

> In maths, powers are also called indices or exponents. The formula $(a^m)^n = a^{mn}$ is known as a law of indices.

② $(a^2)^4 \times (a^5)^5$ equals (   ).

  A. a^{200} B. a^{140} C. a^{33} D. a^{16}

③ $(-a^{n-1})^2$ equals (   ).

  A. a^{2n-1} B. $-a^{2n-1}$ C. a^{2n-2} D. $-a^{2n-2}$

④ $(9 \times 3^n)^2$ equals (   ).

  A. 18×3^{2n} B. 3^{2n+4} C. -18×3^{2n} D. -3^{2n+4}

B. Fill in the blanks

⑤ Calculate: $[(a^3)^2]^6 = $ _____.

⑥ Calculate: $(a^3)^2 \times a^6 = $ _____.

⑦ Calculate: $(a^3)^2 + a^6 = $ _____.

⑧ Calculate: $(a^2 \times a^3)^6 = $ _____.

⑨ Calculate: $(a-b)^2 \times [(b-a)^3]^3 = $ _____.

10 Calculate: $2^{n-3} \times (8 \times 2^{n-2})^3 = $ _____ .

11 Calculate: $-\{-[-(-a^2)^3]^4\}^5 = $ _____ .

C. Questions that require solutions

12 Calculate: $[(x-y)^3]^m \times [(y-x) \times (x-y)^m]^5$.

13 Calculate: $(a^2b^6)^n + 5(-a^nb^{3n})^2 - 3[(-ab^3)^2]^n$.

14 Given that $2^{2x+3} - 2^{2x+1} = 192$, find the value of x.

15 Given that $5^m = a$ and $5^n = b$, use a and b to express 5^{2m} and $5^{2m} + 5^{3n} + 5^{2m+3n}$ respectively.

3.2　Power of a product

Learning objective

Understand and use the meaning of the power of a product and the related laws of indices.

A. Multiple choice questions

1　In these calculations, the correction one is (　　).

A. $(a \times b)^m = a \times b^m$

B. $(a \times b)^m = a^m \times b$

C. $(a \times b)^m = a^m \times b^m$

D. $(a \times b)^m = am \times bm$

> The formulae $(a \times b)^m = a^m \times b^m$ and $\left(\dfrac{a}{b}\right)^m = \dfrac{a^m}{b^m}$ are also called laws of indices; the other laws of indices you have learned include $a^m \times a^n = a^{m+n}$ and $\dfrac{a^m}{a^n} = a^{m-n}$, where m and n are all integers.

2　Of these calculations, the correct one is (　　).

A. $x^3 + x^3 = x^6$

B. $(-a^3)^2 = a^6$

C. $(-a^2)^3 = a^6$

D. $(-a)^3 \times a^5 = a^8$

3　Of these calculations, the incorrect one is (　　).

A. $(-m^3)^2 \times (-n^2)^3 = m^6 n^6$

B. $[(-m^3)^2 \times (-n^2)^3]^3 = -m^{18} n^{18}$

C. $(-m^3 n)^2 \times (-mn^2)^3 = -m^9 n^8$

D. $(-m^2 n)^3 \times (-mn^2)^3 = m^9 n^9$

4　The result of calculating $\left(-\dfrac{2}{3}\right)^{2017} \times \left(\dfrac{3}{2}\right)^{2018}$ is (　　).

A. $\dfrac{2}{3}$

B. $\dfrac{3}{2}$

C. $-\dfrac{3}{2}$

D. $-\dfrac{2}{3}$

B. Fill in the blanks

5　$32a^5 b^{15} = ($＿＿＿＿＿＿$)^5$.

6　$2^{10} \times 8^{10} = ($＿＿＿＿＿＿$)^{10} = 2^{(\text{＿＿＿})}$.

7　Calculate: $\left(-\dfrac{3}{4}ab^2\right)^3 = $＿＿＿＿＿＿.

8 Calculate: $(-3x^3)^2 - (-3x^2)^3 =$ _____.

9 Calculate: $(-0.3)^{2017} \times \left(-\dfrac{10}{3}\right)^{2018} =$ _____.

10 Calculate: $(2^m a^n b^2)^p =$ _____.

11 Given that $(2^4 \times 25^n)^3 = (2^3 \times 5^{n+2})^4$, then $n =$ _____.

C. Questions that require solutions

12 Calculate: $(-x)^2 \times x^3 \times (-2y)^3 + (2xy)^2 \times (-x)^3 \times y$.

13 Calculate: $(a^3 b^9)^n + 2(-a^n b^{3n})^3 + (a^2 b^3)^{3n} + 3(-2a^{2n} b^{3n})^3$.

14 Given that $a^{m+n} = 68$ and $a^m = 17$, find the value of a^{3n}.

15 Calculate: $\left(\dfrac{1}{99} \times \dfrac{1}{98} \times \cdots \times \dfrac{1}{3} \times \dfrac{1}{2} \times 1\right)^{99} \times (1 \times 2 \times 3 \times \cdots \times 98 \times 99 \times 100)^{99}$.

3.3 Factorisation using formulae (1): difference of two squares

 Learning objective

Understand the formula for the difference of two squares; use the formula for factorisation.

 A. Multiple choice questions

1. Of these expressions, (   ) can be factorised by using a formula.

   A. $x^2 - xy$ 　　　　　　　　B. $x^2 + xy$

   C. $x^2 - y^2$ 　　　　　　　　D. $x^2 + y^2$

2. Of these expressions, (   ) cannot be factorised by using the difference of two squares formula.

   A. $-a^2 + b^2$ 　　　　　　　B. $-x^2 - y^2$

   C. $m^2 - n^2$ 　　　　　　　D. $4x^2 - 9y^2$

3. The correct result of factorising $3x^4 - 48$ is (   ).

   A. $3(x + 4)(x - 4)$ 　　　　　B. $3(x^2 + 4)(x^2 - 4)$

   C. $3(x^2 + 4)(x + 4)(x - 4)$ 　D. $3(x^2 + 4)(x + 2)(x - 2)$

 B. Fill in the blanks

4. Factorise: $x^2 - 9 = $ _____ .

5. Factorise: $4x^2 - \dfrac{1}{9}y^2 = $ _____ .

6. Factorise: $-a^2 + \dfrac{1}{9}b^2 = $ _____ .

7. Factorise: $3x^3 - 27x = $ _____ .

8. Factorise: $-4 + a^4 = $ _____ .

9 Factorise: $16x^2 - 49y^4 =$ _____ .

10 Factorise: $(x^2 - y^2) - (x - y)^2 =$ _____ .

C. Questions that require solutions

11 Factorise: $2x - 8x^3$.

12 Factorise: $(a + b)^2 - 100$.

13 Factorise: $7m^3n - 28n^3m$.

14 Factorise: $3m(x - y)^3 + 27m^3(y - x)$.

3.4 Factorisation using formulae (2): completing the square

Learning objective

Understand the formula for completing the square; use the formula for factorisation.

A. Multiple choice questions

1 Of these polynomials, ( ) is a perfect square.

A. $1 - 4m + 2m^2$ 　　　　　　　B. $a^2 + 2a + 4$

C. $4x^2 - 12xy + 9y^2$ 　　　　　D. $a^2 + 4a - 4$

2 Given that $4x^2 + kx + 9$ is a perfect square, then k is ( ).

A. 12 　　　　　　　　　　　　B. 18 or -18

C. 12 or -12 　　　　　　　　D. 6 or -6

3 Of these expressions, ( ) can be factorised by using the perfect square formula.

A. $m^2 + mn + n^2$ 　　　　　　B. $x^2 + 2xy + \frac{1}{4}y^2$

C. $4x^2 - 4xy + y^2$ 　　　　　　D. $4m^2 + mn + n^2$

B. Fill in the blanks

4 Factorise: $x^2 + x + \frac{1}{4}$ = _____.

5 Factorise: $\frac{1}{4}x^2 - x + 1$ = _____.

6 Factorise: $4x^2 - 4x + 1$ = _____.

7 Factorise: $x^2y^2 - 4xy + 4$ = _____.

8 Factorise: $9 - 12t + 4t^2$ = _____.

9 Using factorisation: $($ _____ $)^2 + 20xy + 25y^2 = ($ _____ $)^2.$

10 Using factorisation: _____ $+ 4mn + 1 = ($ _____ $)^2.$

C. Questions that require solutions

11 Factorise: $9x^2 + 42xy + 49y^2.$

12 Factorise: $a^2 - 2a(b + c) + (b + c)^2.$

13 Factorise: $-a^4b^2 + a^2bx - \dfrac{1}{4}x^2.$

14 Factorise: $4 - 12(x - y) + 9(x - y)^2.$

3.5 Factorisation using formulae (3): revision and practice

 Learning objective

Use the formulae for the difference of two squares and completing the square for factorisation.

 A. Multiple choice questions

1 The result of factorising $(x + y)^2 + 6(x + y) + 9$ is (　　).

A. $(x + y - 3)^2$ B. $(x + y + 3)^2$ C. $(x - y + 3)^2$ D. $(x - y - 3)^2$

2 Given that the polynomial $25a^2 + kab + 16b^2$ is a perfect square, then k is (　　).

A. 40   B. ±40   C. 20   D. ±20

3 The result of completely factorising $x^4 - 1$ is (　　).

A. $(x^2 + 1)(x^2 - 1)$ B. $(x^2 - 1)^2$

C. $(x^2 + 1)(x + 1)(x - 1)$ D. $(x + 1)^2(x - 1)^2$

 B. Fill in the blanks

4 $m^2 + mn + (\underline{\hspace{2cm}}) = (\underline{\hspace{2cm}})^2.$

5 Factorise: $16a^2 - \dfrac{1}{9}b^2 = \underline{\hspace{3cm}}.$

6 Factorise: $\dfrac{1}{4}y^2 + y + 1 = \underline{\hspace{3cm}}.$

7 Factorise: $(x + y)^2 + 6(x + y) + 9 = \underline{\hspace{3cm}}.$

8 Factorise: $(x^2 + 2x)^2 + 2(x^2 + 2x) + 1 = \underline{\hspace{3cm}}.$

9 Factorise: $x^3 - 6x^2 + 9x = \underline{\hspace{3cm}}.$

10 Factorise: $x^2 + 2x + 1 - y^2 =$ _____.

C. Questions that require solutions

11 Factorise: $a^2 b^2 - 8abc + 16c^2$.

12 Factorise: $16x^4 - 72x^2 y^2 + 81y^4$.

13 Given that $a - b = \dfrac{1}{2}$ and $ab = \dfrac{1}{8}$, find the value of $-2a^2 b^2 + ab^3 + a^3 b$.

14 Study these equations: $1 - 9 = -8$, $4 - 16 = -12$, $9 - 25 = -16$, $16 - 36 = -20 \cdots$

(a) Use an equation in n (where n is a positive integer) to express a pattern that can be applied to all the above equations.

(b) Use the pattern described in part (a) to write the 10th equation.

3.6　Factorisation using the cross-multiplication method

Learning objective

Understand and apply the cross-multiplication method for factorisation.

A. Multiple choice questions

1 Of these expressions, (　　) can be factorised by using the cross-multiplication method.

A. $x^2 - 3x + 2$　　B. $x^2 - 2x + 4$　　C. $x^2 - 3x - 2$　　D. $x^2 + x + 1$

2 Of these expressions, (　　) cannot be factorised by using the cross-multiplication method.

A. $x^2 + x - 2$　　B. $x^2 - 7x + 12$　　C. $x^2 - 4x - 12$　　D. $x^2 - x + 12$

3 The result of factorising $x^2 - 4x - 5$ should be (　　).

A. $(x-1)(x+5)$　　B. $(x+1)(x-5)$　　C. $(x+1)(x+5)$　　D. $(x-1)(x-5)$

B. Fill in the blanks

4 $x^2 +$ _____ $+ 20 = (x+4)($ _____ $)$.

5 $x^2 -$ _____ $- 20 = (x+4)($ _____ $)$.

6 Given that $x^2 + 3x + 2 = (x+1)(x+m)$, then $m =$ _____ .

7 Given that $x^2 + px + 8 = (x-2)(x-q)$, then $p =$ _____ , $q =$ _____ .

8 Factorise: $x^2 - 5x + 6 =$ _____ .

9 Factorise: $x^2 + x - 12 =$ _____ .

10 Factorise: $x^2 - x - 6 =$ _____ .

C. Questions that require solutions

11 Factorise: $x^4 - 5x^2 + 4$.

12 Factorise: $(x + y)^2 - 4(x + y) - 12$.

13 Factorise: $x^2y^2 - 10xy + 16$.

14 Given that the polynomial $x^2 + ax - 6$ can be factorised into a product of two linear factors, each with integral coefficients, find the value of a.

3.7 Factorisation by grouping

Learning objective

Understand how to factorise by grouping terms.

A. Multiple choice questions

1 Of these polynomials, (   ) cannot be factorised by grouping.

A. $5x + mx + 5y + my$

B. $5x + mx + 3y + my$

C. $5x - mx + 5y - my$

D. $5x - mx + 10y - 2my$

2 To factorise $4 - x^2 + 2x^3 - x^4$, the reasonable way of grouping is (   ).

A. $(4 - x^2) + (2x^3 - x^4)$

B. $(4 - x^2 - x^4) + 2x^3$

C. $(4 - x^4) + (-x^2 + 2x^3)$

D. $(4 - x^2 + 2x^3) - x^4$

B. Fill in the blanks

3 Factorise: $ab + b^2 - ac - bc$ = (_____) − ($ac + bc$) = (_____)(_____).

4 Factorise: $ax^2 + ax - b - bx$ = ($ax^2 - bx$) + (_____) = (_____)(_____).

5 Factorise: $2ax + 4bx - ay - 2by$ = (_____) + (_____)
= (_____)(_____).

6 Factorise: $x^2 - a^2 - 2ab - b^2$ = (_____) − (_____)
= (_____)(_____).

7 Factorise: $a^2 + ax - b^2 + bx$ = _____ .

8 Factorise: $xy - x - y + 1$ = _____ .

C. Questions that require solutions

9 Factorise: $x^2 + 4y - 1 - 4y^2$.

10 Factorise: $x^2 - 2x - a^2 - 2a$.

11 Factorise: $7x^2 - 3y + xy - 21x$.

12 Factorise: $4a^2 - 4(ab + 4) + b^2$.

13 Factorise: $x^2 - 4xy + 4y^2 - 2x + 4y - 3$.

14 Factorise: $\dfrac{1}{4} - n^2 + m^2 - m$.

Unit test 3

A. Multiple choice questions

1 Of these calculations, the correction one is ( ).

A. $(2a^m)^3 = 2a^{m+3}$ B. $(2a^m)^3 = 8a^{m+3}$ C. $(2a^m)^3 = 2a^{3m}$ D. $(2a^m)^3 = 8a^{3m}$

2 $(a^3 b^5)^2$ equals ( ).

A. $a^5 b^7$ B. $a^3 b^7$ C. $a^6 b^7$ D. $a^6 b^{10}$

3 $(-x^n)^3$ equals ( ).

A. $-3x^n$ B. $3x^n$ C. $-x^{3n}$ D. x^{3n}

4 Of these expressions, ( ) can be factorised by using a formula.

A. $x^2 - 10$ B. $x^2 + 10$ C. $x^2 - 100$ D. $x^2 + 100$

5 Of these expressions, ( ) is the incorrect one.

A. $(a \times b)^m = a^m \times b^m$ B. $(a^m \times a^n) = a^{mn}$

C. $(a \div b)^m = a^m \div b^m$ D. $(a^m \div a^n) = a^{m-n}$

6 Of these expressions, ( ) cannot be factorised by using the cross-multiplication method.

A. $x^2 - 8x - 7$ B. $x^2 - 8x + 7$ C. $x^2 - 8x + 12$ D. $x^2 - 8x - 9$

B. Fill in the blanks

7 Calculate: $[(m^2)^2]^2 = $ _____.

8 Calculate: $(x^3)^2 \times (x^2)^3 = $ _____.

9 Calculate: $\left(-\dfrac{2}{3}xy^3\right)^3 = $ _____.

10 Calculate: $(-2ax^2)^4 - (-3a^2x^4)^2 = $ _____.

11 $16x^4y^8 = ($ _____ $)^4$.

12 Factorise: $ax^2 - 25a = $ _____ .

13 Factorise: $xy^2 - 20xy + 100x = $ _____ .

14 Factorise: $x^3 - 2x^2 - 3x = $ _____ .

15 Factorise: $-x^2 + a^2 - 2a + 1 = $ _____ .

D. Questions that require solutions

16 Given that $(m^3 \times 16^n)^2 = (m^2 \times 4^{n+2})^3$, find the value of n.

17 Factorise: $(a - b)(3a + b)^2 + (a + 3b)^2(b - a)$.

18 Factorise: $ax - 2x + 3a - 6$.

19 Calculate: $(-5)^{2017} \times (-2)^{2018} + 10^{2017}$. (Express your result in power form.)

20 Given that $a + b = 10$ and $ab = 15$, find the value of $a^2 + b^2 + ab^2 + a^2b + 3ab$.

21 First factorise the algebraic expression $2a(x^2 - 2)^2 - 2ax^2$, where $a \neq 0$, and then find the value(s) of x for which the value of the algebraic expression is zero.

Chapter 4　Algebraic fractions

4.1　The meaning of algebraic fractions

 Learning objective

Understand and manipulate algebraic fractions.

 A. Multiple choice questions

1 Among the algebraic expressions $\dfrac{a^2}{2}$, $\dfrac{1}{a+b}$, $\dfrac{a}{x-1}$, $\dfrac{x^2}{x}$, $-m^2$ and $\dfrac{x+y}{x}$, there are (　　) algebraic fractions.

A. 2　　　　　　　　　　　　　　　B. 3

C. 4　　　　　　　　　　　　　　　D. 5

2 Given that the value of the algebraic expression $\dfrac{x^2-4}{x+2}$ is zero, then the value of x is (　　).

A. 2　　　　　　　　　　　　　　　B. -2

C. 0　　　　　　　　　　　　　　　D. ±2

3 To make the algebraic fraction $\dfrac{x+1}{x-3}$ valid, the range of values of x is (　　).

A. $x \neq 0$　　　　　　　　　　　B. $x \neq -1$

C. $x \neq 3$　　　　　　　　　　　D. $x \neq -3$

4 For the algebraic fraction $\dfrac{x+a}{3x-1}$, when $x=-a$, the correct conclusion is (　　).

A. The value of the algebraic fraction is zero.

B. The algebraic fraction is not valid.

C. If $a \neq -\dfrac{1}{3}$, the value of the algebraic fraction is zero.

D. If $a \neq \dfrac{1}{3}$, the value of the algebraic fraction is zero.

B. Fill in the blanks

5 Choose some digits and letters from "1, 2, a, b and c" to form two algebraic expressions, one of them being a polynomial expression and the other an algebraic fraction. The polynomial expression you formed is _____ and the algebraic fraction is _____. (Just write one of each.)

6 Rewrite each of these algebraic fractions as a division.

(a) $\dfrac{a}{2a - b} =$ _____

(b) $-\dfrac{a - b}{2x - y} =$ _____

7 Rewrite each of these expressions as an algebraic fraction.

(a) $(a + 2) \div 3(b - 5) =$ _____

(b) $-(x + 1) \div x =$ _____

8 Adele needs a days to finish a project on her own while Becky needs b days to finish it on her own. If Adele and Becky work together, then they need _____ days to finish the work.

9 When $x = -1$ and $y = 2$, then the value of $\dfrac{x^2 - 2xy}{y + 2}$ is _____.

10 Given that the algebraic fraction $\dfrac{x - 1}{x^2 - 4}$ is not meaningful, then the value of x is _____.

11 When $x =$ _____, the value of the algebraic fraction $\dfrac{x^2 - 2x}{x^2 + 2x - 3}$ is 1.

C. Questions that require solutions

12 Calculate the value of each of these algebraic fractions when $x = 2$ and $y = -1$.

(a) $\dfrac{x^2 + y^2}{x + y}$

(b) $\dfrac{x^2 - y}{2y + 2}$

13 Given that the value of the algebraic fraction $\dfrac{6x^2 - 5x - 6}{3x + 2}$ is 0, find the value of x.

14 For what values of x is the value of the algebraic fraction $\dfrac{x - 2}{3 + 2x}$ negative?

15 For what integer values of m is the value of the algebraic fraction $\dfrac{2m + 7}{m - 1}$ a positive integer?

4.2　Properties of algebraic fractions

Learning objective

Understand the properties of algebraic fractions; apply them for simplifying.

A. Multiple choice questions

1 Of these algebraic fractions, the one equal to the value of $-\dfrac{1+a}{a-3}$ is (　　).

A. $-\dfrac{a+1}{3+a}$　　　　B. $\dfrac{a-1}{a-3}$　　　　C. $\dfrac{-1-a}{3-a}$　　　　D. $\dfrac{-1-a}{a-3}$

2 If the values of x and y in the algebraic fraction $\dfrac{x+y}{xy}$ are simultaneously doubled, then

the value of the algebraic fraction is (　　).

A. doubled　　　　　　　　　　　B. unchanged

C. halved　　　　　　　　　　　　D. none of the above

3 Of these algebraic fractions, the simplest is (　　).

A. $\dfrac{y^2}{x^3yz^2}$　　　　B. $\dfrac{x+1}{1-x^2}$　　　　C. $\dfrac{2x^2-4}{6x+12}$　　　　D. $\dfrac{4x-1}{4x}$

4 Study these equations.

① $\dfrac{-(a-b)}{c} = -\dfrac{a-b}{c}$　　　　　　② $\dfrac{-x+y}{-x} = \dfrac{x-y}{x}$

③ $\dfrac{-a+b}{c} = -\dfrac{a+b}{c}$　　　　　　④ $\dfrac{-m-n}{m} = -\dfrac{m-n}{m}$

(　　) are true.

A. ① and ②　　　B. ③ and ④　　　C. ① and ③　　　D. ② and ④

B. Fill in the blanks

5 Write a suitable numerator or denominator in each equation, to make it true.

(a) $\dfrac{x}{3y} = \dfrac{(\quad)}{6xy^2}$　　　(b) $\dfrac{2x^2y}{xy^3} = \dfrac{(\quad)}{y^2}$　　　(c) $\dfrac{-2x}{1-2x} = \dfrac{(\quad)}{2x^2-x}$

(d) $\dfrac{x-y}{5y} = \dfrac{(y-x)^2}{(\quad)}$　　　(e) $\dfrac{3x^2-xy}{9x^2-6xy+y^2} = \dfrac{x}{(\quad)} = \dfrac{(\quad)}{9x^2-y^2}$

6 Among the algebraic fractions $\dfrac{x}{3a}$, $\dfrac{x-y}{x^2-y^2}$, $\dfrac{ab}{a^2-b^2}$ and $\dfrac{a+b}{a-b}$, there are _____ simplest algebraic fractions.

7 Given that the value of the algebraic fraction $\dfrac{2m+4}{m^2-4}$ is a positive integer, then the value that the integer m can take is _____ .

8 Given that $2x+y=0$ and $xy\neq 0$, then the value of the algebraic fraction $\dfrac{x^2+2xy}{xy+y^2}$ is _____ .

9 When $1<x<2$, the value of the algebraic fraction $\dfrac{\sqrt{(x-2)^2}}{x-2}+\dfrac{\sqrt{(x-1)^2}}{x-1}$ is _____ .

C. Questions that require solutions

10 Simplify these algebraic fractions.

(a) $\dfrac{-15xy^2z^{15}}{5xyz^5}$

(b) $\dfrac{a^2+6a-7}{a^2-4a+3}$

(c) $\dfrac{a^{n+2}-a^2b^n}{a^{2n+1}-ab^{2n}}$

(d) $\dfrac{-(x+y)(x-y)^2}{(y+x)(y^2-x^2)}$

(e) $\dfrac{a^2+b^2-c^2+2ab}{a^2-b^2-c^2-2bc}$

11 Given that x, y and z satisfy $\dfrac{y+z}{x}=\dfrac{z+x}{y}=\dfrac{x+y}{z}=k$, find the value of k.

4.3 Multiplying and dividing algebraic fractions (1)

Learning objective

Learn how to multiply and divide algebraic fractions.

A. Multiple choice questions

1 The result of calculating $\dfrac{4a^2}{3b} \times \left(-\dfrac{b^2}{2a}\right)$ is (　　).

A. $\dfrac{4a^2 - b^2}{6ab}$　　　　B. $\dfrac{2}{3}ab$　　　　C. $-\dfrac{2}{3}ab$　　　　D. $-2ab$

2 Of the four calculations:

① $\dfrac{a}{y} \times \dfrac{x}{b}$　　　　② $\dfrac{n}{m} \times \dfrac{2m}{n}$　　　　③ $\dfrac{4}{x} \div \dfrac{2}{x}$　　　　④ $\dfrac{a}{b^2} \div \dfrac{2a^2}{b^2}$

the results of (　　) are algebraic fractions.

A. ① and ④　　　　B. ① and ③　　　　C. ② and ④　　　　D. ③ and ④

3 The result of calculating $-3xy \div \dfrac{2y^2}{3x}$ is (　　).

A. $-2y^3$　　　　B. $-2y$　　　　C. $\dfrac{9x^2}{2y}$　　　　D. $-\dfrac{9x^2}{2y}$

4 Of these equations, (　　) is true.

A. $\dfrac{-x - y}{x} = -\dfrac{x - y}{x}$

B. $x \div \dfrac{y}{x - y} = \dfrac{x}{y} \div (x - y)$

C. $(a + b) \div \dfrac{a - b}{ab} = \dfrac{a^2 - b^2}{ab}$

D. $\dfrac{x^2 - y^2}{x - y} \div \dfrac{x + y}{x - y} = x - y$

B. Fill in the blanks

5 When multiplying two algebraic fractions, use the product of the numerators as the _____ and the product of the denominators as the _____. When dividing one algebraic fraction by another, first turn the fraction that is the divisor _____, then _____ the dividend by this new fraction.

6 Calculate: $\dfrac{ab^2}{2c^2} \div \dfrac{-3a^2b^2}{4c} = $ _____.

7 Calculate: $\dfrac{3a - 3b}{10ab} \times \dfrac{50a^2b^2}{a^2 - b^2} = $ _____.

8 Calculate: $\dfrac{a^2 - 1}{a} \div \dfrac{(1 - a)^2}{-a^2} = $ _____.

9 Calculate: $(x - 2) \div \dfrac{(x - 2)^2}{x^2} \times x^3 = $ _____.

10 If $x - \dfrac{1}{x} = 2$, then $x^2 + \dfrac{1}{x^2} = $ _____.

C. Questions that require solutions

11 Calculate: $\dfrac{x^2 - 4y^2}{x^2 + y^2 + 2xy} \div \dfrac{2y + x}{x^2 + xy}$.

12 Calculate: $\dfrac{2x - 6}{x^2 - 4x + 4} \div \dfrac{12 - 4x}{x^2 + x - 6} \times \dfrac{1}{x + 3}$.

13 First simplify, and then evaluate: $\dfrac{a - 1}{a + 2} \times \dfrac{a^2 - 4}{a^2 - 1} \div \dfrac{1}{(a + 1)^2}$, where $a^2 = a - 1$.

14 Given that $\dfrac{a^2 + 1}{a} = 5$, find the value of $\dfrac{a^4 + a^2 + 1}{a^2}$.

4.4　Multiplying and dividing algebraic fractions (2)

Learning objective

Multiply and divide algebraic fractions by using the laws of indices or factorising.

A. Multiple choice questions

1 The result of calculating $\left(-\dfrac{b^2}{3a}\right)^2$ is (　　).

 A. $-\dfrac{b^4}{6a^2}$ B. $\dfrac{b^4}{6a^2}$ C. $-\dfrac{b^4}{9a^2}$ D. $\dfrac{b^4}{9a^2}$

2 The result of calculating $\left(-\dfrac{3ax^2}{2y}\right)^3$ is (　　).

 A. $-\dfrac{27a^3x^6}{8y^3}$ B. $\dfrac{27a^3x^6}{8y^3}$ C. $-\dfrac{9a^3x^6}{6y^3}$ D. $\dfrac{9a^3x^6}{6y^3}$

3 Of these equations, (　　) is correct.

 A. $\left(\dfrac{x+y}{xy}\right)^2 = \dfrac{x^2+y^2}{x^2y^2}$ B. $\left(-\dfrac{y^2}{2x}\right)^3 = -\dfrac{y^5}{6x^3}$

 C. $\left(\dfrac{x^2}{y}\right)^n \times \left(\dfrac{y^n}{x^{n+1}}\right)^2 = \dfrac{y^2}{x^2}$ D. $\left(\dfrac{n}{3m}\right)^2 \div \left(-\dfrac{n}{m}\right)^6 = \dfrac{m^4}{9n^4}$

4 The result of calculating $\left(-\dfrac{b}{a^2}\right)^7 \times \left(-\dfrac{a}{b^3}\right)^2$ is (　　).

 A. $-\dfrac{1}{a^{12}b^2}$ B. $\dfrac{b}{a^{12}}$ C. $-\dfrac{b}{a^{12}}$ D. $\dfrac{1}{a^{12}b^2}$

B. Fill in the blanks

5 Calculate: $\left(-\dfrac{3ax^3}{4pm^3}\right)^2 = $ _____.

6 Calculate: $(2xy^2)^2 \div \left(-\dfrac{4y^3}{3x}\right) = $ _____.

7 Given that n is a positive integer, then simplifying $\left(-\dfrac{a}{b}\right)^{2n+1} \times \left(-\dfrac{b}{a}\right)^{2n}$ gives

_____.

8 Calculate: $\left(\dfrac{y}{x}\right)^2 \div \left(-\dfrac{x^3}{y}\right)^3 \times \left(\dfrac{5y}{2x^2}\right)^2 =$ _____.

9 Calculate: $\dfrac{(x^2-y^2)^2}{(x^2+y^2)^3} \div \left(\dfrac{x+y}{x^2+y^2}\right)^3 =$ _____.

C. Questions that require solutions

10 Calculate: $\left(-\dfrac{x}{y^2}\right)^2 \times \left[-\left(\dfrac{y}{x}\right)^2\right]^3 \div \left(\dfrac{-y}{x}\right)^3$

11 Calculate: $\dfrac{x^2-5x+6}{x^2-x-2} \div \dfrac{x^2-7x+12}{x^2-2x-8}$.

12 Calculate: $\left(\dfrac{x-1}{x^2-x-2}\right)^2 \div \dfrac{x^2-2x+1}{2-x} \div \left(\dfrac{1}{x^2+x}\right)^2$.

13 First simplify, and then evaluate:

$\left[\dfrac{-m^3n}{(m-n)^2}\right]^4 \times \left(\dfrac{n^2-mn}{m}\right)^3 \div m^4n^{10} \times \left(\dfrac{mn-n^2}{m}\right)^6$, where $m=1$ and $n=-1$.

4.5 Adding and subtracting algebraic fractions (1)

Learning objective

Learn how to add and subtract algebraic fractions with the same denominator.

A. Multiple choice questions

1 The result of calculating $\dfrac{a + 3b}{3ab} - \dfrac{a + 9b}{3ab}$ is (　　).

A. $\dfrac{2}{a}$　　　　　B. $-\dfrac{2}{a}$　　　　　C. $\dfrac{4}{a}$　　　　　D. $-\dfrac{4}{a}$

2 The result of calculating $\dfrac{a^2}{a - b} - \dfrac{b^2}{b - a}$ is (　　).

A. $\dfrac{a^2 + b^2}{a - b}$　　　　B. $a + b$　　　　C. $a - b$　　　　D. $b - a$

3 The result of calculating $\dfrac{a}{a^2 - b^2} - \dfrac{b}{b^2 - a^2}$ is (　　).

A. $\dfrac{1}{a + b}$　　　B. $\dfrac{1}{a - b}$　　　C. $\dfrac{a - b}{a^2 - b^2}$　　　D. $\dfrac{a + b}{a^2 - b^2}$

4 Of these equations, (　　) is correct.

A. $\dfrac{1}{a} - \dfrac{1}{b} = \dfrac{1}{a - b}$　　　　　　B. $\dfrac{y}{x} - \dfrac{y - 1}{x} = \dfrac{1}{x}$

C. $\dfrac{x}{x^2 - y^2} - \dfrac{y}{x^2 - y^2} = \dfrac{x - y}{x^2 - y^2}$　　　　D. $\dfrac{a}{x - y} - \dfrac{b - a}{y - x} = \dfrac{2a - b}{x - y}$

B. Fill in the blanks

5 Calculate: $\dfrac{2}{3a} - \dfrac{5}{3a} =$ _____.

6 Calculate: $\dfrac{3}{x^2 - 9} - \dfrac{x}{x^2 - 9} =$ _____.

7 Calculate: $\dfrac{x+2}{x-2} + \dfrac{x-4}{2-x} =$ _____.

8 Calculate: $\dfrac{x}{(x-y)^3} + \dfrac{y}{(y-x)^3} =$ _____.

9 Calculate: $\dfrac{x^2}{(x-y)^2} - \dfrac{y^2}{(y-x)^2} =$ _____.

10 Given that $\dfrac{a+9b}{3ab} + \dfrac{A}{3ab} = \dfrac{2}{a}$, then $A =$ _____.

C. Questions that require solutions

11 Calculate: $\dfrac{x^2+1}{x^2-5x+6} + \dfrac{1-2x}{x^2-5x+6} - \dfrac{5}{x^2-5x+6}$.

12 Calculate: $\dfrac{x^2-a^2}{x^2-9a^2} + \dfrac{(x-2a)^2}{9a^2-x^2} + \dfrac{5a^2-4ax}{x^2-9a^2}$.

13 Calculate: $\dfrac{2c+a}{(a-b)(b-c)(c-a)} + \dfrac{b+c}{(a-b)(c-b)(c-a)} - \dfrac{b-a-c}{(b-a)(c-b)(a-c)}$.

4.6 Adding and subtracting algebraic fractions (2)

Learning objective

Learn how to add and subtract algebraic fractions with different denominators

A. Multiple choice questions

1. The simplest common denominator of the algebraic fractions $\dfrac{4a}{3bc}$, $\dfrac{3b}{4ac}$ and $-\dfrac{5c}{6ab}$ is (   ).

   A. $72a^2b^2c^2$　　　　B. $12a^2b^2c^2$　　　　C. $72abc$　　　　D. $12abc$

2. The simplest common denominator of the algebraic fractions $\dfrac{m+2}{m-2}$ and $\dfrac{-5}{m^2-4}$ is (   ).

   A. $m-2$

   B. m^2-4

   C. $m+2$

   D. $(m+2)(m^2-4)$

3. After reducing fractions to a common denominator, if the denominator of the algebraic fraction $\dfrac{3a}{a^2-b^2}$ is changed to $2(a-b)^2(a+b)$, then the numerator should be changed to (   ).

   A. $6a(a-b)^2(a+b)$

   B. $2(a-b)$

   C. $6a(a-b)$

   D. $6a(a+b)$

B. Fill in the blanks

4. Calculate: $\dfrac{1}{a}+\dfrac{1}{b}=$ _____.

5. Calculate: $\dfrac{1}{2a^2}-\dfrac{b}{a}=$ _____.

6. Calculate: $\dfrac{1}{2a-b}-\dfrac{1}{2a+b}=$ _____.

7. Calculate: $\dfrac{x^2}{x+y}-x+y=$ _____.

8 Calculate: $1 - \dfrac{1}{a} + \dfrac{2}{a-1} =$ _____.

9 If $\dfrac{A}{x-2} + \dfrac{B}{x+1} = \dfrac{3}{(x-2)(x+1)}$, then $A =$ _____ and $B =$ _____.

C. Questions that require solutions

10 Calculate: $-\dfrac{1}{x+3} - \dfrac{6}{9-x^2} - \dfrac{x-1}{6+2x}$.

11 Calculate: $\dfrac{7}{2x-4} - \dfrac{3}{x+2} + \dfrac{12}{4-x^2}$.

12 Calculate: $\dfrac{a^2}{a^2-4} + \dfrac{4}{a+2-a^2} + \dfrac{4a-4}{a^4-5a^2+4}$.

13 First simplify, and then evaluate: $\dfrac{x}{1+x} - \dfrac{1}{1-x} - \dfrac{x^3-2x+1}{x^2-1}$, where $x = -2$.

14 Calculate: $\dfrac{x+2}{x+1} - \dfrac{x+3}{x+2} - \dfrac{x-4}{x-3} + \dfrac{x-5}{x-4}$.

4.7 Adding and subtracting algebraic fractions (3)

Learning objective

Add and subtract algebraic fractions, and apply this to solving related problems.

A. Multiple choice questions

1. Of these statements, the correct one is (   ).

   A. The simplest common denominator of $\dfrac{2}{3a^2}$ and $\dfrac{1}{6ab^2}$ is $18a^3b^2$.

   B. The simplest common denominator of $\dfrac{x}{a(x-y)}$ and $\dfrac{y}{b(y-x)}$ is $ab(x-y)(y-x)$.

   C. The simplest common denominator of $\dfrac{4}{3x}$, $\dfrac{x-1}{-2x^2}$ and $\dfrac{x+1}{4x^3}$ is $-12x^6$.

   D. The simplest common denominator of $\dfrac{1}{x+1}$, $\dfrac{1}{1+2x+x^2}$ and $\dfrac{-2}{1-x^2}$ is $(x+1)^2(x-1)$.

2. Of these calculations with algebraic fractions, the correct one is (   ).

   A. $\dfrac{a-b}{c} - \dfrac{a+b}{c} = \dfrac{a-b-a+b}{c} = 0$

   B. $\dfrac{ad}{bc} \div \dfrac{dy}{ax} \times \dfrac{bx}{ay} = \dfrac{a^3}{b^2c}$

   C. $\dfrac{5}{a-b} - \dfrac{1}{(b-a)^2} = \dfrac{5a-5b+1}{(a-b)^2}$

   D. $x - y + \dfrac{2y^2}{x+y} = \dfrac{x^2+y^2}{x+y}$

3. The result of this calculation with algebraic fractions $\left(\dfrac{x}{x-2} - \dfrac{x}{2+x}\right) \div \dfrac{4x}{2-x}$ is (   ).

   A. $-\dfrac{1}{x+2}$

   B. $\dfrac{1}{x+2}$

   C. $-\dfrac{1}{x-2}$

   D. $\dfrac{1}{x+2}$

B. Fill in the blanks

4. Calculate: $\dfrac{c}{2ab} + \dfrac{b}{3a^2c} = $ _____

5. Calculate: $2x + 2 + \dfrac{5}{x-1} = $ _____

6 Calculate: $\dfrac{a-1}{a} \div \left(a - \dfrac{1}{a}\right) =$ _____

7 Calculate: $\left(\dfrac{x}{x-y} + \dfrac{y}{y-x}\right) \div \dfrac{xy}{x-y} =$ _____

8 Given $\dfrac{3x-5}{x^2-2x-3} = \dfrac{A}{x-3} + \dfrac{B}{x+1}$, then $A =$ _____ and $B =$ _____

C. Questions that require solutions

9 Calculate: $\left(\dfrac{x-1}{x+1} - \dfrac{x+1}{x+2}\right) \div \dfrac{x+3}{x^2+4x+4}$.

10 Calculate: $\left(\dfrac{x^3-x^2}{x^2+x} - \dfrac{1-x^2}{x+1}\right) \div \dfrac{2x+1}{x+1}$.

11 Calculate: $\left(\dfrac{a+1}{2a-2} - \dfrac{3}{2a^2-2} - \dfrac{a+3}{2a+2}\right) \times \dfrac{4a^2-4}{3}$.

12 First simplify, and then evaluate: $\left(\dfrac{a-2}{a^2+2a} - \dfrac{a-1}{a^2+4a+4}\right) \div \dfrac{a-4}{a+2}$, where a satisfies $a^2+2a-1 = 0$.

13 Given that the algebraic fraction $\dfrac{6x^2+2x+4}{x(x-1)(x+2)} = \dfrac{A}{x} + \dfrac{B}{x-1} + \dfrac{C}{x+2}$, find the values of A, B and C.

4.8　Algebraic fraction equations that can be transformed to linear equations in one variable

Learning objective

Learn how to transform algebraic fractions into linear equations in one variable and apply this to solve equations.

A. Multiple choice questions

1 Of these equations, (　　) are algebraic fraction equations.

① $\dfrac{x+1}{3} = 1$ 　　② $\dfrac{3}{x+1} = 4$ 　　③ $\dfrac{x^2-1}{x+1} = 1$ 　　④ $\dfrac{x}{2} + \dfrac{x-1}{3} = 2$

⑤ $\dfrac{x+1}{\pi} + 2 = x$

A. 1 　　　　　　B. 2 　　　　　　C. 3 　　　　　　D. 4

2 Of these equations, (　　) has the solution $a = -1$.

A. $\dfrac{2}{a+1} - \dfrac{1}{a+2} = 0$ 　　　　　　B. $\dfrac{2}{a-1} - \dfrac{2}{a} = 0$

C. $\dfrac{2}{a-1} + \dfrac{1}{a+2} = 0$ 　　　　　　D. $\dfrac{2}{a} = 1 - \dfrac{2}{a+2}$

3 The root(s) of the equation $\dfrac{(x-2)(x+3)}{x^2-4} = 0$ is/are (　　).

A. -3 　　　　　B. 2 　　　　　C. 2 or -3 　　　　　D. -2 or 3

B. Fill in the blanks

4 When $x = $ _____, the value of the algebraic fraction $\dfrac{x-5}{3x+1}$ equals 0.

5 When $x = $ _____, the value of the algebraic fraction $\dfrac{2x-5}{x-2}$ equals -1.

6 The solution to the equation $\dfrac{x}{x-1} = \dfrac{2}{1-x}$ is _____.

7 Given that $x = 1$ is a root of the equation $\dfrac{5}{ax - 3} = 1$, then $a =$ _____.

8 If the solutions to the equations $\dfrac{1}{x - 2} = 2$ and $\dfrac{x}{x + a} = -\dfrac{3}{2}$ in x are equal, then the value of a is _____.

9 Given that there is no real solution to the equation $\dfrac{2 + x}{x - 3} = \dfrac{1 - m}{3 - x}$ in x, then the value of m is _____.

C. Questions that require solutions

10 Solve the equation: $\dfrac{1}{3x - 6} = \dfrac{3}{4x - 8}$.

11 Solve the equation: $\dfrac{x}{3 + x} - \dfrac{x}{2 - x} = 2$.

12 Solve the equation: $\dfrac{2}{2x - 5} - \dfrac{1}{x - 5} = \dfrac{2}{2x - 1} - \dfrac{1}{x - 3}$.

13 Solve the equation: $\dfrac{1}{x + 2} + \dfrac{1}{x + 7} = \dfrac{1}{x + 3} + \dfrac{1}{x + 6}$

4.9 Integer exponents and their operations (1)

Learning objective

Understand and recognise integer exponents and their operations.

A. Multiple choice questions

1. Of these calculations, the correct one is (   ).

   A. $(-2)^{-2} = 4$ B. $-2^{-2} = \dfrac{1}{4}$ C. $5x^{-2} = \dfrac{1}{5x^2}$ D. $2(xy)^{-1} = \dfrac{2}{xy}$

2. Of these calculations, the correct one is (   ).

   A. $a^2 \times (a^{-3})^2 = a^{-3}$ B. $(a-2)^{-2} = \dfrac{1}{a^2 - 4}$

   C. $a^2 \div a^{-6} = a^{-4}$ D. $(a^{-2})^{-3} \div a^{-2} = a^8$

3. Arranging -4^{-2}, -0.2^2, $\left(-2\dfrac{2}{3}\right)^0$ and $\left(\dfrac{3}{5}\right)^{-3}$ in order, according to their values,

   leads to (   ).

   A. $-0.2^2 < -4^{-2} < \left(\dfrac{3}{5}\right)^{-3} < \left(-2\dfrac{2}{3}\right)^0$ B. $-0.2^2 < -4^{-2} < \left(-2\dfrac{2}{3}\right)^0 < \left(\dfrac{3}{5}\right)^{-3}$

   C. $-4^{-2} < -0.2^2 < \left(\dfrac{3}{5}\right)^{-3} < \left(-2\dfrac{2}{3}\right)^0$ D. $-4^{-2} < -0.2^2 < \left(-2\dfrac{2}{3}\right)^0 < \left(\dfrac{3}{5}\right)^{-3}$

B. Fill in the blanks

4. Convert these to expressions with negative and positive exponents.

   (a) $\dfrac{3}{x^2 y^3} = $ _____

   (b) $\dfrac{2}{x} + \dfrac{3}{y^2} = $ _____

   (c) $\dfrac{(x-y)^3}{4(x+y)^5} = $ _____

   (d) $\dfrac{2x-y}{x^5 y} = $ _____

5 Convert these expressions to expressions without negative exponents.

(a) $5x^{-2} =$ _____

(b) $\dfrac{3^{-2}xy^{-4}}{5a^{-3}b^{-5}} =$ _____

(c) $(x^{-1} + y^{-1})^{-1} =$ _____

(d) $\dfrac{-3x(x + y)^{-2}}{y^{-2}} =$ _____

6 Calculate: $\left(x - \dfrac{1}{2}\right)^0 =$ _____.

7 Calculate: $2^{-2} + (-2)^2 =$ _____.

8 Calculate: $\left(\dfrac{2}{3}\right)^{-2} - \left(\dfrac{4}{5}\right)^{-1} + \left(\dfrac{6}{7}\right)^0 =$ _____.

9 Calculate: $\dfrac{x^{-1} - y^{-1}}{x^{-1} + y^{-1}} =$ _____.

C. Questions that require solutions

10 Calculate: $(-2)^{-3} - 2^{-3} + \left(\dfrac{1}{2}\right)^{-3} - \left(-\dfrac{1}{2}\right)^{-3}$.

11 Calculate: $\left(-\dfrac{1}{2}\right)^{-2} - 2^3 \times 0.125 + 2006^0 + 1 - 11$.

12 Calculate: $(x^{-2} - y^{-2}) \div (x^{-1} + y^{-1}) + \dfrac{1}{y} - \dfrac{1}{x}$.

13 Given that $x^2 - y^2 = xy$ and $xy \neq 0$, find the value of $x^2 y^{-2} + x^{-2} y^2$.

4.10 Integer exponents and their operations (2)

Learning objective

Understand and recognise standard form, and convert numbers to and from standard form.

A. Multiple choice questions

1 Of these expressions the one written correctly in standard form is (   ).

  A. 0.99×10^{-7} B. $(9 \times 10)^{-2}$ C. 32×10^{-5} D. 1.5×10^{-1}

2 Given that $0.000\,082 = 8.2 \times 10^{n}$, then n is (   ).

  A. 4   B. 5   C. -4 D. -5

3 Of these equations using standard form to express numbers, (   ) is correct.

  A. $-3.515 \times 10^{5} = -35\,150$ B. $1 \times 10^{-5} = 0.000\,001$

  C. $-1.41 \times 10^{-4} = -0.000\,141$ D. $-3.05 \times 10^{-3} = -0.000\,305$

B. Fill in the blanks

4 Write each of these numbers in standard form.

  (a) $-0.5768 = $ _____

  (b) $-100.109 = $ _____

  (c) $0.043\% = $ _____

  (d) $\dfrac{367}{100\,000} = $ _____

  (e) $3\,679\,900 = $ _____

  (f) $0.001\,237 = $ _____ (to the nearest hundred thousandth)

  (g) $0.00\,006\,283 = $ _____ (to 2 significant figures)

5 These numbers are expressed in standard form. Write them as normal numbers.

  (a) $5.03 \times 10^{6} = $ _____ ;

  (b) $-3.15 \times 10^{-4} = $ _____ ;

  (c) $1.09 \times 10^{-5} = $ _____ ;

  (d) $-4.23 \times 10^{-3} = $ _____ ;

6 Calculate: $(-0.125)^{-101} \div \left(-\dfrac{1}{8}\right)^{-100} = $ _____ .

7 Given that $x^{2} + x^{-2} = 5$, then $x^{4} + x^{-4} = $ _____ .

C. Questions that require solutions

8 Calculate: $(-2.4 \times 10^{-6}) \times (2.5 \times 10^{3}) \times (8 \times 10^{-7})$.

9 Calculate: $(a^{-1} + b^{-1})^{-1} \div (a^{-2} - b^{-2})^{-1}$.

10 Calculate: $\left(\dfrac{2a - b}{a + b}\right)^{-2} - \dfrac{a}{b - 2a} + \left(\dfrac{2a - b}{2b}\right)^{-1}$.

11 Given that $x + \dfrac{1}{x} = 2$, find $\dfrac{x^{2048} + x^{-2048} - 2}{x^{2013} + x^{-2013}}$.

12 Given that $10^{-2\alpha} = 3$ and $10^{-\beta} = -\dfrac{1}{5}$, find the value of $10^{6\alpha + 2\beta}$.

Unit test 4

A. Multiple choice questions

1 Among $\dfrac{x-y}{2}$, $\dfrac{x+3}{x}$, $\dfrac{5+x}{\pi}$ and $\dfrac{a+b}{a-b}$, there are (   ) algebraic fractions.

    A. 1   B. 2   C. 3   D. 4

2 Of these algebraic fractions, the one that has the same value as $\dfrac{x-2}{x+3}$ is (   ).

    A. $\dfrac{(x-2)(x+2)}{(x+3)(x-3)}(x \neq 3)$
           B. $\dfrac{(x-2)(x-2)}{(x+3)(x+3)}$

    C. $\dfrac{(x-2)(x-3)}{(x+3)(x-3)}(x \neq 3)$
           D. $\dfrac{(x-2)(x+3)}{(x+3)(x-2)}(x \neq 2)$

3 If $a = (-2)^{-2}$, $b = (-2)^{0}$ and $c = \left(-\dfrac{1}{2}\right)^{-1}$, then the relationship between a, b and c

    is (   ).

    A. $a > c > b$ B. $b > a > c$ C. $a > b > c$ D. $c > a > b$

4 Given that $\dfrac{5x+1}{(x-1)(x-2)} = \dfrac{a}{x-1} + \dfrac{11}{x-2}$, then the value of a is (   ).

    A. -6 B. -3 C. 3   D. 6

5 The distance between A and B is m metres. A messenger planned to take t hours to walk from A to B to deliver a message. Due to an unexpected change in plans, he now must reach B in h hours. Therefore, he should walk (   ) more metres per hour.

    A. $\dfrac{t-h}{m}$ B. $\dfrac{ht}{mt-mh}$ C. $\dfrac{ht}{mh-mt}$ D. $\dfrac{mt-mh}{ht}$

6 Given that one root of the equation $\dfrac{x}{x^2+m} = \dfrac{2}{x-3m}$ is $x = 1$, then the value of m is

    (   ).

    A. $\dfrac{1}{5}$ B. $-\dfrac{1}{4}$ C. $-\dfrac{1}{2}$ D. $-\dfrac{1}{5}$

B. Fill in the blanks

7 The simplest common denominator of the algebraic fractions $\dfrac{2a}{a-1}$, $\dfrac{1}{1-a^2}$ and $-\dfrac{1}{a+1}$ is _____ .

8 When x is _____ , the algebraic fraction $\dfrac{x-1}{2+x}$ is valid.

9 When x is _____ , the value of the algebraic fraction $\dfrac{x^2-1}{x+1}$ is 0.

10 Biologists found that the diameter of a kind of bacterium is $0.000\,47$ m. In standard form, this is written as _____ m.

11 Calculate: $\dfrac{1}{3}ab \div \left(-\dfrac{2b^2}{3}\right) =$ _____ .

12 Calculate: $\dfrac{1}{x-2} + \dfrac{1}{x^2-5x+6} =$ _____ .

13 Given that $a - \dfrac{1}{a} = 3$, then $a^2 + \dfrac{1}{a^2} =$ _____ .

14 Given that $a = 2b$, then the value of $\dfrac{a^2-b^2}{ab}$ is _____ .

15 Calculate: $(-2)^{-2} - 2^{-2} =$ _____ .

16 Use negative exponents to convert the given expression to an expression without a denominator.

$\left(-\dfrac{3^{-1}a^{-2}b^3}{2a^2b^{-2}}\right)^{-1} =$ _____

17 If $\dfrac{1}{x} - \dfrac{1}{y} = 2$, the algebraic expression $\dfrac{4(y-x)+xy}{x-y} =$ _____ .

18 Solving the algebraic fraction equation $\dfrac{x-2}{x+2} - \dfrac{16}{x^2-4} = 1$, the result is _____.

19 Given that $x^2 - 4x + 1 = 0$, then $\dfrac{x^2}{x^4 - 4x^2 + 1} = $ _____.

C. Short answer questions

20 Calculate: $\left(-\dfrac{1}{2a}\right) \div \left(\dfrac{2a}{b}\right)^2 \times \dfrac{3b^2}{a^2} \div (-2a^2b)^3$.

21 Calculate: $\dfrac{2}{2x+3} - \dfrac{3}{2x-3} + \dfrac{2x+15}{4x^2-9}$.

22 Calculate: $\dfrac{1}{2m} - \dfrac{1}{m-n} \times \left(\dfrac{m-n}{2m} - m + n\right)$.

23 Calculate: $\left(-2\dfrac{1}{3}\right)^{-1} \times \left(-\dfrac{6}{7}\right)^{-2} + \left(-\dfrac{2}{3}\right)^{-2} \times \left(-\dfrac{1}{3}\right)^0$.

24 Solve the equation: $\dfrac{1}{x-4} - \dfrac{1}{x-3} = \dfrac{1}{x-2} - \dfrac{1}{x-1}$.

D. Questions that require solutions

25 First simplify, and then evaluate:

$$\left(x - \dfrac{x}{x+1}\right) \times \dfrac{x+1}{x^2+3x+2} \div \dfrac{x^2-2x}{x^2-4}, \text{ where } x = -\dfrac{1}{2}.$$

26 Given that $\dfrac{-2x+7}{3x^2+17x-6} = \dfrac{A}{x+6} + \dfrac{B}{3x-1}$, find the values of A and B.

27 A courier was asked to take a message from his base camp to a sentry post 3 kilometres away and then return immediately, along the same route. It took him 50 minutes to complete the whole journey. If his speed travelling to the post was 1.5 times his speed as he returned, find the speed at which he returned.

28 Use three methods to convert the algebraic fraction $\dfrac{3x-4}{x^2-3x+2}$ into the sum of several algebraic expressions that have no x-term in their numerators.

Chapter 5　Real numbers, fractional indices and surds

5.1　Operations with real numbers (1)

Learning objective

Understand and manipulate expressions involving surds and indices.

A. Multiple choice questions

1 Of these statements, (　　) is incorrect.

    A.　For any non-zero real number a and positive integer n, $a^{-n} = \dfrac{1}{a^n}$.

    B.　For any non-zero real number a, $a^0 = 1$.

    C.　For any non-zero real number a, $a^0 = 0$.

    D.　For any non-zero real numbers a and b and positive integer n, $(ab)^{-n} = \dfrac{1}{a^n} \times \dfrac{1}{b^n}$.

2 Of these equations, (　　) is correct.

    A.　$\sqrt{4.9} = 0.7$　　　B.　$\sqrt[3]{-2} = -\sqrt[3]{2}$　　　C.　$\sqrt{(-2)^2} = -2$　　D.　$\sqrt{4} = \pm 2$

3 Of these equations, (　　) is correct.

    A.　$5\sqrt{2} - 2\sqrt{2} = 3$　　B.　$5 - 2\sqrt{2} = 3\sqrt{2}$　　C.　$\sqrt{21} \div \sqrt{3} = \sqrt{7}$　　D.　$\sqrt{7} - \sqrt{2} = \sqrt{5}$

4 The expression that is equal to $a\sqrt{-\dfrac{1}{a}}$ is (　　).

    A.　$\sqrt{a}$　　　　　　　B.　$-\sqrt{a}$　　　　　　C.　$\sqrt{-a}$　　　　　　D.　$-\sqrt{-a}$

B. Fill in the blanks

5 Calculate: $5\sqrt{2} - 3\sqrt{2} + \sqrt{2} = $ _____ .

6 Calculate: $5\sqrt{2} - 3\sqrt{2} + \sqrt{3} = $ _____ .

7 Calculate: $5\sqrt{21} \times 2\sqrt{3} =$ _____ .

8 Calculate: $(\sqrt{3} \times \sqrt{10} - 3\sqrt{30}) \div \sqrt{6} =$ _____ .

9 Calculate $(\sqrt{3} \times \sqrt{10} - 3\sqrt{60}) \div \sqrt{6} =$ _____ .

10 Calculate: $\sqrt[3]{-27} + \sqrt[3]{0.008} - \sqrt[3]{\dfrac{1}{64}} =$ _____ .

11 Calculate: $\dfrac{\sqrt{27}}{\sqrt{3}} - \sqrt{2} \times \sqrt{32} =$ _____ .

C. Questions that require solutions

12 Calculate: $\left(\dfrac{1}{\sqrt{3} - \sqrt{2}}\right)^{-2} - 2(\sqrt{3} + \sqrt{2})^2 + (\sqrt{3} - \sqrt{2})^0$.

13 Calculate:

(a) $(\sqrt{9 + 4\sqrt{5}} - \sqrt{9 - 4\sqrt{5}})^2$

(b) $\sqrt{9 + 4\sqrt{5}} - \sqrt{9 - 4\sqrt{5}}$.

14 (a) Given that $x = \sqrt{2017} - 1$, find the value of $x^2 + 2x + 2$.

(b) Given that $x = \sqrt{2015} - 2014$, find the value of $x^2 + 4028x + 2014 \times 2015$.

15 The diagram shows a large square with side length $\sqrt{3} + \sqrt{2}$, with a small square with side length $\sqrt{3} - \sqrt{2}$ cut from one corner. Find the area of the shaded region (correct to 3 s. f.).

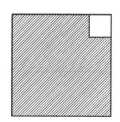

Diagram for question 15

5.2 Operations with real numbers (2)

Learning objective
Understand and evaluate expressions with surds and indices, both exactly and with a calculator.

A. Multiple choice questions

1 In order for $\sqrt{-a}$ to be an integer, a must be (   ).

A. an integer

B. a perfect square number

C. the product of −1 and a perfect square number

D. a negative real number

2 Of these numbers, the rational number between π and 4 is (   ).

A. 3.1   B. 3.2   C. $\dfrac{\pi+3}{2}$ D. $\dfrac{\pi+4}{2}$

3 Based on the diagram, the result of $\sqrt{(a-b)^2} + \sqrt{(a+b)^2} - \sqrt{a^2}$ is (   ).

A. $3a$ B. a

C. $-3a$ D. $-a$

Diagram for question 3

B. Fill in the blanks

4 Use a calculator to evaluate: $\sqrt{11} - \sqrt[3]{12} \approx$ _____. (correct to 0.01).

5 Use a calculator to evaluate: $(\sqrt{19} - 5)^2 \approx$ _____. (correct to 0.01).

6 Use a calculator to evaluate: $\sqrt{20.13} - \sqrt[3]{201.4} \approx$ _____. (correct to 3 s.f.).

7 Use a calculator to evaluate: $\dfrac{\sqrt{3.14}}{5} \cdot \sqrt[3]{0.618} \approx$ _____. (correct to 3 s.f.).

8 If the square roots of $x - 11$ are ±4, then the cube root of x is _____.

9 Calculate: $(3\sqrt{2} - \sqrt{17})^{2017} \cdot (3\sqrt{2} + \sqrt{17})^{2018} =$ _____. *

10 Given three numbers 1, $\sqrt{3}$ and 3, think of one more number so that the four numbers together form a proportion: _____ : _____ = _____ : _____. (You only need to write one such proportion.)

C. Questions that require solutions

11 Without using a calculator, evaluate:

(a) $\sqrt{(\sqrt{2} - \sqrt{5})^2} - \sqrt{(\sqrt{6} - \sqrt{5})^2}$

(b) $\sqrt{\left(-\dfrac{1}{2}\right)^{-4} + (\sqrt{3})^4}$.

12 Given that $0 < a < 2$, simplify $\sqrt{a^2 - 4a + 4} + \sqrt{a^2 - 2a + 1}$.

13 The diagram shows a window. The upper part is a semicircle and the lower part is a square. Given that the area of the window is 3 square metres, find the width, x, of the window (correct to 0.01).

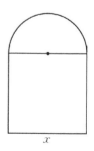

Diagram for question 13

---

* The dot stands for 'multiply'.

14 As shown in the diagram, the large square is formed by four congruent right-angled triangles with the lengths of the two non-hypotenuse sides being $\sqrt{2}$ and $\sqrt{7}$ respectively. They enclose a small shaded square.

(a) Find the area of the small shaded square.

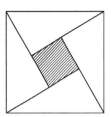

(b) Find the length of one side of the large square.

Diagram for question 14

5.3 Fractional indices (1)

Learning objective

Understand and calculate with fractional indices.

A. Multiple choice questions

1 Of these calculations, (　　) is correct.

　A. $\sqrt[8]{(-5)^2} = (-5)^{\frac{1}{4}}$

　B. $\sqrt[4]{16 \times 5^3} = 4 \times 5^{\frac{3}{4}}$

　C. $\sqrt[3]{13^2 - 5^2} = 13^{\frac{2}{3}} - 5^{\frac{2}{3}}$

　D. $\sqrt{\dfrac{3}{5}} = \left(\dfrac{5}{3}\right)^{-\frac{1}{4}}$

2 Of these calculations, (　　) is correct.

　A. $\left(\dfrac{1}{5}\right)^{-2} = \sqrt{5}$

　B. $\left(\dfrac{1}{5}\right)^{-2} = -25$

　C. $\left(\dfrac{1}{5}\right)^{\frac{1}{2}} = \dfrac{\sqrt{5}}{5}$

　D. $5^{-\frac{1}{2}} \times 5^2 = -5$

3 Of these calculations, (　　) is correct.

　A. $5^3 \times 5^{\frac{1}{3}} = 5$　　B. $5^{-1} \times 5^{\frac{1}{2}} = \dfrac{\sqrt{5}}{5}$　　C. $\left(5^{\frac{1}{2}}\right)^{-\frac{1}{2}} = 1$　　D. $5^{-\frac{1}{2}} \times 5^{\frac{1}{2}} = 5$

B. Fill in the blanks

4 Rewrite these roots in terms of indices, where p and q are positive integers and $q > 1$.

　(a) $\sqrt[q]{a^p} = $ _____ $(a \geqslant 0)$

　(b) $\dfrac{1}{\sqrt[q]{a^p}} = $ _____ $(a > 0)$

5 Rewrite these roots in terms of indices.

　(a) $\sqrt[5]{2^3} = $ _____

　(b) $\sqrt[3]{5^2} = $ _____

　(c) $\dfrac{1}{\sqrt{3^5}} = $ _____ .

6 Express in root form:

　(a) $3^{\frac{4}{5}} = $ _____

　(b) $5^{\frac{2}{3}} = $ _____

　(c) $2^{-\frac{3}{5}} = $ _____ .

78

7 Calculate: $(27 \times 64)^{\frac{1}{3}} =$ _____.

8 Calculate: $\left(2\frac{41}{64}\right)^{\frac{1}{2}} + \left(2\frac{10}{27}\right)^{-\frac{2}{3}} =$ _____.

9 Use a calculator to find the approximate value: $\sqrt[3]{25} \times \sqrt{5} \approx$ _____ (correct to 3 d.p.)

10 Use a calculator to find the approximate value: $\sqrt{35} \div \sqrt[3]{5} \approx$ _____ (correct to 3 s.f.).

C. Questions that require solutions

11 Calculate. (Give your answers in index form.)

(a) $5^{\frac{2}{3}} \times 5^{\frac{1}{4}}$　　　　(b) $6^3 \div 6^{\frac{2}{3}}$　　　　(c) $(7^{\frac{2}{3}})^{-\frac{5}{4}}$

12 Calculate. (Give your answers in index form.)

(a) $5^{\frac{1}{3}} \times 25^{\frac{2}{5}}$　　　　(b) $36^{\frac{1}{3}} \div 6^{\frac{1}{2}}$　　　　(c) $(49^{\frac{1}{3}})^{-2} \times 7^{\frac{1}{2}}$

13 Calculate.

(a) $\left(\frac{1}{8}\right)^{-\frac{1}{3}} + \left(-\frac{1}{4}\right)^{-2} - 8^{\frac{1}{2}} \times \left(\frac{1}{4}\right)^{\frac{1}{2}}$　　　　(b) $(169^{\frac{1}{2}} - 81^{\frac{1}{2}})^{-\frac{1}{2}} + 196^{\frac{1}{3}} \times 49^{\frac{1}{6}} \times 2^{\frac{1}{3}}$.

14 Given that $y = \sqrt{x-8} + \sqrt{8-x} - \frac{4}{3}$, find the value of x^y.

5.4   Fractional indices (2)

Learning objective

Calculate with fractional indices and surds.

A. Multiple choice questions

1 Of these calculations, the correct one is (   ).

    A. $3a^{-2} = \dfrac{1}{3a^2}$
                    B. $\sqrt[5]{a^3} \div \sqrt[5]{a^2} = \sqrt[5]{a}$

    C. $(a^{-\frac{1}{3}})^3 = -a$
                     D. $\sqrt[6]{(-5)^2} = \sqrt[3]{-5}$

2 $\sqrt[4]{\sqrt[3]{a}}$ ($a \geqslant 0$) can be simplified as (   ).

    A. $a^{\frac{13}{12}}$
           B. $a^{\frac{7}{12}}$
           C. $a^{\frac{1}{12}}$
           D. $a^{\frac{1}{24}}$

3 The result of simplifying $\sqrt{-a} \cdot \sqrt[3]{a^2}$ is (   ). *

    A. $a\sqrt[6]{a}$
           B. $a\sqrt[6]{-a}$
           C. $-a\sqrt[6]{-a}$
           D. $-a\sqrt[6]{a}$

B. Fill in the blanks

4 Given that $\sqrt[5]{3 \cdot \sqrt{27}} = 3^n$, then n = _____.

5 Given that $\dfrac{1}{\sqrt[3]{4^2}} = 2^n$, then n = _____.

6 Calculate: $\sqrt{2} \times \sqrt[3]{4} \times \sqrt[4]{8}$ = _____.

7 Calculate: $\dfrac{\sqrt{3} \times \sqrt[3]{9}}{\sqrt[4]{27}}$ = _____.

8 Calculate: $(a^{\frac{1}{2}} + a^{-\frac{1}{2}})^2 \times (a^{\frac{1}{2}} - a^{-\frac{1}{2}})^2$ = _____.

---

  * In these questions, the dot stands for 'multiply'.

80

9 Calculate: $(a - b) \div (a^{\frac{1}{2}} - b^{\frac{1}{2}})$ = _____.

10 Calculate: $\dfrac{a - 3a^{\frac{1}{2}}b^{\frac{1}{2}} + 2b}{a^{\frac{1}{2}} - b^{\frac{1}{2}}}$ = _____.

C. Questions that require solutions

11 Calculate $(29^{\frac{1}{2}} - 13^{\frac{1}{2}})^{\frac{1}{4}} \times (29^{\frac{1}{2}} + 13^{\frac{1}{2}})^{\frac{1}{4}} \times (29^{\frac{1}{3}} - 13^{\frac{1}{2}})^{0}$.

12 The diagram shows square $ABCD$ with area 5 square units and square $EFGC$ with area 3 square units.

(a) Find the perimeter of the rectangle $BHFE$.

(b) Find the area of the rectangle $BHFE$.

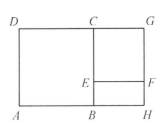

Diagram for question 12

13 Given that $a^{\frac{1}{2}} - a^{-\frac{1}{2}} = -2$, find:

(a) the value of $a + a^{-1}$ (b) the value of $a^2 + a^{-2}$.

14 Given that $3^a = \sqrt[3]{2}$ and $3^b = \sqrt[5]{4}$, find the value of 3^{4a+3b}.

5.5 Surds (1)

Learning objective

Understand and manipulate algebraic expressions involving surds.

A. Multiple choice questions

1 Of these numbers, ( ) is a surd.

    A. 8   B. $\sqrt{8}$ C. $\sqrt[3]{8}$ D. $\sqrt[4]{16}$

2 Of these statements, the incorrect one is ( ).

    A. A surd is an irrational number.

    B. A surd is a real number.

    C. A surd is usually expressed in nth root form such as a square root or a cube root.

    D. Any number expressed as a square root is a surd.

> A surd expressed as an nth root is known as an nth order surd. A second order or third order surd is commonly known as a **quadratic surd** or **cubic surd.**

B. Fill in the blanks *

3 To make $\sqrt{-2x}$ meaningful, the set of values that x can take is _____ .

4 To make $\dfrac{\sqrt{x}}{2x-1}$ meaningful, the set of values that x can take is _____ .

5 If $\sqrt{a^2} = a$, then a _____ ; if $\sqrt{a^2} = -a$, then a _____ .

6 When $x \leqslant 1$, $\sqrt{x^2 - 2x + 1} = $ _____ .

7 The result of simplifying $\sqrt{(a-2)^2} + (\sqrt{2-a})^2$ is _____ .

8 Given that $\sqrt{x-1} + \sqrt{1-x} = y + 4$, then the square root(s) of x^y is/are _____ .

---

* From now on, we mainly study quadratic surds, which are also simply called surds.

9 If $\sqrt{(x-5)^2} + 2\sqrt{y+2} = 0$, then the value of $x - y$ is _____.

10 Observe the equations: $\sqrt{1 + \dfrac{1}{3}} = 2\sqrt{\dfrac{1}{3}}$, $\sqrt{2 + \dfrac{1}{4}} = 3\sqrt{\dfrac{1}{4}}$, $\sqrt{3 + \dfrac{1}{5}} = 4\sqrt{\dfrac{1}{5}}$, $\cdots$

What pattern can you find? Use an equation with the natural number $n(n \geqslant 1)$ to describe it: _____.

C. Questions that require solutions

11 Calculate.

(a) $\sqrt{(3.14 - \pi)^2}$ 　(b) $-(-\sqrt{3^2})^2$ 　(c) $\sqrt{\left[\left(\dfrac{2}{3}\right)^{-1}\right]^2}$ 　(d) $\left(\dfrac{3}{\sqrt{0.5^2}}\right)^2$

12 First simplify $\sqrt{1 - 2x + x^2} - \sqrt{1 + 4x + 4x^2}$, and then find its value when $x = -2$.

13 First simplify $2a - \sqrt{a^2 - 4a + 4}$, and then find its value when $a = \sqrt{2}$.

14 Given that $-1 < a < 1$, simplify $\sqrt{(a + 1)^2} + \sqrt{(a - 1)^2}$.

15 Given that $x < y$, simplify $y - x - \sqrt{(x-y)^2}$.

16 Given that the side lengths of $\triangle ABC$, a, b and c, are all integers and that a and b satisfy $\sqrt{a-2} + b^2 - 6b + 9 = 0$, find the side length c in $\triangle ABC$.

17 Given that real number a satisfies $\sqrt{(3-a)^2} + \sqrt{a-4} = a$, find the value of $a - 3^2$.

5.6 Surds (2)

 Learning objective

Solve problems involving surds.

A. Fill in the blanks

1 The condition to make the equation $\sqrt{xy} = \sqrt{x} \times \sqrt{y}$ true is _____.

2 The condition to make the equation $\sqrt{\dfrac{x}{y}} = \dfrac{\sqrt{x}}{\sqrt{y}}$ true is _____.

3 Simplify $\sqrt{28} = \sqrt{\underline{} \times \underline{}} = \sqrt{\underline{}} \times \sqrt{\underline{}} = \underline{}$.

4 The condition to make the equation $\sqrt{x^2 - 1} = \sqrt{x - 1} \times \sqrt{x + 1}$ true is _____.

5 The condition to make the equation $\sqrt{\dfrac{x}{x - 2}} = \dfrac{\sqrt{x}}{\sqrt{x - 2}}$ true is _____.

6 Of the quadratic surds $\sqrt{7}$, $\sqrt{27}$, $\sqrt{30}$, $\sqrt{\dfrac{1}{2}}$ and $\sqrt{210}$, those that can be simplified further are _____.

7 Given that $m < 0$, $n \geqslant 0$, simplify $\sqrt{m^2 n} =$ _____.

8 Simplify these quadratic surds.

(a) $\sqrt{8} =$ _____

(b) $\sqrt{76} =$ _____

(c) $\sqrt{\dfrac{16}{25}} =$ _____

(d) $\sqrt{1\dfrac{7}{9}} =$ _____

(e) $\sqrt{96} =$ _____

(f) $\sqrt{\dfrac{4b^2}{9a^2}}\,(a > 0, b \geqslant 0) =$ _____

B. Questions that require solutions

In Questions 9–14, simplify the quadratic surds.

9 $\sqrt{121 \times 0.64}$

10 $\sqrt{\dfrac{81}{196}}$

11 $\sqrt{\dfrac{25\, y^4}{36\, x^2}}\ (x > 0)$

12 $\sqrt{\dfrac{3}{a^2 - 2a + 1}}\ (a < 1)$

13 $\sqrt{\dfrac{15\, x^4 y}{1.25 \times 0.03}}\ (y > 0)$

14 $\sqrt{\left(\dfrac{2}{3}\right)^{2} \times \left(-\dfrac{3}{5}\right)^{2}}$

15 The diagram shows the position of a real number a on a number line. Simplify $\sqrt{(1-a)^{2}} + \sqrt{a^{2}}$.

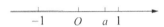

Diagram for question 15

16 Given that the positions of real numbers a and b are shown on the number line, simplify $\sqrt{a^{2}} - \sqrt{b^{2}} - \sqrt{(a-b)^{2}}$.

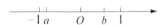

Diagram for question 16

17 Given that n is a positive interger and $\sqrt{135n}$ is an integer, find the smallest possible value of n.

5.7 Surds in their simplest form

 Learning objectives

Simplify algebraic expressions involving surds in simplest form.

 A. Fill in the blanks

1 Of the expressions $\sqrt{5ab}$, $\sqrt{0.1}$, $\sqrt{a^2 - b^2}$,

$\sqrt{a^2 - 2ab + b^2}$ $(a \geqslant b)$, $\sqrt{\dfrac{x}{2}}$ and $\dfrac{\sqrt{x}}{2}$, the

surds in simplest form are _____.

> A quadratic surd is said to be in its simplest form when the number under the square root is an integer without a perfect square as a factor.

2 Simplify each surd into its simplest form.

(a) $\sqrt{12} =$ _____

(b) $\sqrt{18x} =$ _____

(c) $\sqrt{\dfrac{y}{x}} =$ _____ $(x > 0)$

(d) $\sqrt{4\dfrac{1}{2}} =$ _____

3 In each space, write the simplest multiplying factor such that the product of the factor and the quadratic surd is a rational number, for example, $3\sqrt{2}$ and $\sqrt{2}$.

(a) $\sqrt{32}$ and _____

(b) $\sqrt{3a}$ and _____

(c) $\sqrt{3a^2}$ and _____

(d) $\sqrt{3a^3}$ and _____

 B. Questions that require solutions

In Questions 4–9, simplify each expression into its simplest form.

4 $\dfrac{1}{5}\sqrt{3\dfrac{1}{5}}$

5 $\sqrt{40a^2 b^3 c}$ $(a \geqslant 0, b \geqslant 0)$

6 $\sqrt{\dfrac{60m^2 n}{a}}$ $(m \geqslant 0, a > 0)$

7 $\sqrt{x^4 + 3x^2}$ $(x > 0)$

8 $\dfrac{1}{2}\sqrt{16x^4 + 64a^2}$

9 $\sqrt{\dfrac{b}{a^2} + \dfrac{a}{b^2}}$ $(a > 0, b > 0)$

10 Given that $x = 4 - \sqrt{2}$ and $y = 4 + \sqrt{2}$, find the values of the algebraic expressions $\sqrt{x^2 - 2xy + y^2}$ and $xy^2 + x^2 y$.

11 Given that $a = \sqrt{5} + \sqrt{2}$ and $b = \sqrt{5} - \sqrt{2}$, find the value of the algebraic expression $\sqrt{a^2 + b^2 + 7}$.

12 Given that a is a real number, find the value of algebraic expression

$\sqrt{a+2} - \sqrt{8-4a} + \sqrt{-a^2}$.

13 If the product of two algebraic expressions containing quadratic surds does not contain a quadratic surd, then we say that the two algebraic expressions are **rationalising factors** of each other. For example, $\sqrt{a}$ and $\sqrt{a}$, $3 + \sqrt{6}$ and $3 - \sqrt{6}$ are rationalising factors of each other.

Write the rationalising factor of each of these expressions.

(a) $5\sqrt{2}$ and _____

(b) $\sqrt{x - 2y}$ and _____

(c) $\sqrt{x} - \sqrt{2y}$ and _____

(d) $2 + \sqrt{3}$ and _____

(e) $3 + 2\sqrt{2}$ and _____

(f) $3\sqrt{2} - 2\sqrt{3}$ and _____

14 Observe and find a pattern: $\dfrac{1}{\sqrt{2}+1} = \sqrt{2} - 1$, $\dfrac{1}{\sqrt{3}+\sqrt{2}} = \sqrt{3} - \sqrt{2}$, $\dfrac{1}{2+\sqrt{3}} = 2 - \sqrt{3}$.

(a) Now try to simplify:

$\dfrac{1}{\sqrt{11} + \sqrt{10}} =$

$\dfrac{1}{\sqrt{7} + 2\sqrt{2}} =$

(b) Simplify $\dfrac{1}{\sqrt{n} + \sqrt{n+1}}$. (Show your working.)

5.8   Like quadratic surds

Learning objective

Understand and combine like quadratic surds.

A. Fill in the blanks

1   Of the expressions $\frac{1}{3}\sqrt{12a^3x^3}$, $3a\sqrt{\dfrac{x}{3a}}$, $3x\sqrt{\dfrac{a}{x^3}}$ and $\sqrt{4a^3x}$, the like quadratic surds for $\sqrt{3ax}$ are _____.

> Two quadratic surds are said to be like quadratic surds if, in their simplest form, they can be expressed as $a\sqrt{n}$ and $b\sqrt{n}$, where a and b are numbers not in surd form and n is a whole number (or an expression).

2   The like quadratic surds in the set of expressions $2\sqrt{45}$, $\frac{3}{5}\sqrt{125}$, $10\sqrt{0.05}$ and $\frac{1}{8}\sqrt{50}$ are _____.

3   Given that $\sqrt{5x+8}$ and $\sqrt{7}$ are like quadratic surds, then the smallest positive integer of x is _____.

4   Given that the quadratic surds $\sqrt[a+b]{2a-8}$ and $\sqrt{b+5}$ are in their simplest form and are like quadratic surds, then ab = _____.

5   Combine the like quadratic surds: $\sqrt{3} - 3\sqrt{2} - 5\sqrt{3} + \sqrt{2}$ = _____.

6   Combine the like quadratic surds: $\frac{1}{3}\sqrt{a} + 6\sqrt{ab} - 5\sqrt{a} - \frac{3}{4}\sqrt{ab}$ = _____.

7   Combine the like quadratic surds: $3\sqrt{x} - 2y\sqrt{2x} + 6y\sqrt{2x} - \frac{1}{5}\sqrt{x}$ = _____.

8   Given that $x = \sqrt{2} - 1$ and $y = 3\sqrt{2} + 2$, then $2y - 3x$ = _____.

91

B. Questions that require solutions

In Questions 9–15, combine the like quadratic surds.

9 $\sqrt{5} - \sqrt{6} - 2\sqrt{5} + \dfrac{\sqrt{6}}{3} + \dfrac{3}{5}\sqrt{5}$

10 $(8\sqrt{abc} - 6\sqrt{ab}) - \left(7\sqrt{abc} - \dfrac{1}{3}\sqrt{ab} + \dfrac{1}{2}\sqrt{abc}\right)$

11 $9\sqrt{3} + 7\sqrt{12} - 5\sqrt{48}$

12 $\sqrt{12\dfrac{1}{2}} + 4\sqrt{1.75} - \dfrac{1}{6}\sqrt{28} + \sqrt{200}$

13 $(6\sqrt{0.75} - \sqrt{18} - \sqrt{12}) - \left(\dfrac{4}{5}\sqrt{108} - 5\sqrt{2} - 4\sqrt{0.5}\right)$

14 $\dfrac{1}{2}x\sqrt{4x} + 6x\sqrt{\dfrac{x}{9}} - 2x^2\sqrt{\dfrac{1}{x}}$

⑮ $\left(4b\sqrt{\dfrac{a}{b}} + \dfrac{2}{a}\sqrt{a^5 b^3}\right) - 3ab\left(\sqrt{\dfrac{1}{ab}} + \sqrt{4ab}\right)(b > 0)$

16 Given that the quadratic surds $2a\sqrt{3a + 2b}$ and $^{a+b-3}\!\sqrt{3b - a}$ are in their simplest form and are like quadratic surds, find the values of a and b.

5.9 Adding and subtracting quadratic surds

Learning objectives

Know how to add and subtract quadratic surds.

A. Fill in the blanks

1 Given eight quadratic surds: ① $\sqrt{32}$, ② $\sqrt{27}$, ③ $\sqrt{125}$, ④ $4\sqrt{45}$, ⑤ $2\sqrt{8}$, ⑥ $\sqrt{18}$, ⑦ $\sqrt{12}$ and ⑧ $\sqrt{15}$, after they are simplified, _____ have the same radicands as $\sqrt{2}$, _____ have the same radicand as $\sqrt{3}$, and _____ have the same radicand as $\sqrt{5}$. (Fill in the code numbers.)

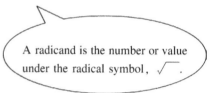

A radicand is the number or value under the radical symbol, $\sqrt{}$.

2 Of these equations, the correct one is _____. (Fill in the code number.)

① $3\sqrt{3} + 3 = 6\sqrt{3}$

② $\frac{1}{7}\sqrt{7} = 1$

③ $\sqrt{2} + \sqrt{6} = \sqrt{8} = 2\sqrt{2}$

④ $\frac{\sqrt{24}}{\sqrt{3}} = 2\sqrt{2}$

3 Calculate: $3\sqrt{8} - 5\sqrt{32} = $ _____.

4 Calculate: $\sqrt{27} - \frac{1}{3}\sqrt{18} - \sqrt{12} = $ _____.

5 Calculate: $(\pi + 1)^0 - \sqrt{12} + \sqrt{3} = $ _____.

6 Calculate: $3\sqrt{48} - 9\sqrt{\frac{1}{3}} + 3\sqrt{12} = $ _____.

7 Calculate: $(\sqrt{48} + \sqrt{20}) - (\sqrt{12} - \sqrt{5}) = $ _____.

8 Calculate: $\left(\sqrt{80} - \sqrt{1\frac{4}{5}}\right) - \left(\sqrt{3\frac{1}{5}} + \frac{4}{5}\sqrt{45}\right) = $ _____.

B. Questions that require solutions

For Questions 9–14, complete the calculation.

9 $\dfrac{1}{2}(\sqrt{2}+\sqrt{3}) - \dfrac{3}{4}(\sqrt{2}-\sqrt{27})$

10 $\left(\sqrt{12}-4\sqrt{\dfrac{1}{8}}\right) - \left(3\sqrt{\dfrac{1}{3}}-4\sqrt{0.5}\right)$

11 $3\sqrt{2x}-5\sqrt{8x}+7\sqrt{18x}$

12 $\dfrac{2}{3}\sqrt{9x}+6\sqrt{\dfrac{x}{4}}-2x\sqrt{\dfrac{1}{x}}$

13 $a\sqrt{\dfrac{1}{a}}+\sqrt{4b}-\dfrac{\sqrt{a}}{2}+b\sqrt{\dfrac{1}{b}}$

14 $2a\sqrt{\dfrac{b}{a}} - b\sqrt{\dfrac{a}{b}} + \dfrac{1}{a}\sqrt{a^3b} - \dfrac{2}{b}\sqrt{a\,b^3}$

15 Simplify first and then evaluate: $x\sqrt{\dfrac{1}{x}} + \sqrt{4y} - \dfrac{\sqrt{x}}{2} + \dfrac{\sqrt{y^3}}{y}$, where $x = 4$ and $y = \dfrac{1}{9}$.

16 Simplify first and then evaluate: $\left(6x\sqrt{\dfrac{y}{x}} + \dfrac{3}{y}\sqrt{x\,y^3}\right) - \left(4y\sqrt{\dfrac{x}{y}} + \sqrt{36xy}\right)$, where $x = \dfrac{3}{2}$ and $y = 27$.

17 Simplify first and then evaluate: $\dfrac{a^2 - 2ab + b^2}{a^2 - b^2} \div \left(\dfrac{1}{a} - \dfrac{1}{b}\right)$, where $a = \sqrt{2} + 1$ and $b = \sqrt{2} - 1$.

18 Given that $4x^2 + y^2 - 4x - 6y + 10 = 0$, evaluate $\left(\dfrac{2}{3}x\sqrt{9x} + y^2\sqrt{\dfrac{x}{y^3}}\right) - \left(x^2\sqrt{\dfrac{1}{x}} - 5x\sqrt{\dfrac{y}{x}}\right)$.

5.10 Multiplying and dividing quadratic surds

Learning objectives

Know how to multiply and divide quadratic surds.

A. Fill in the blanks (All the letters represent positive numbers.)

1 $\sqrt{2} \times \sqrt{3} =$ _____

2 $\sqrt{24} \times \sqrt{54} =$ _____

3 $\sqrt{15} \times \sqrt{\dfrac{5}{3}} =$ _____

4 $2\sqrt{5a} \times \sqrt{10a} =$ _____

5 $4\sqrt{5} \div \left(-5\sqrt{1\dfrac{4}{5}}\right) =$ _____

6 $6\sqrt{27xy} \div \sqrt{\dfrac{x}{y}} =$ _____

7 $\sqrt{\dfrac{3xy}{7}} \times \left(-\dfrac{1}{2}\sqrt{28x^2 y}\right) =$ _____

8 $\sqrt{18mn} \times \sqrt{2m^2 n^4} =$ _____

B. Short-answer questions

9 $\sqrt{18} \times \sqrt{24} \times \sqrt{27}$

10 $\sqrt{xy} \times \sqrt{x^3 y} \times \sqrt{x\,y^2}$

11 $\sqrt{3\dfrac{1}{3}} \div \left(\dfrac{2}{5}\sqrt{2\dfrac{1}{3}}\right) \times \left(4\sqrt{1\dfrac{2}{5}}\right)$

12 $-\dfrac{4}{3}\sqrt{18} \div \left(2\sqrt{8} \times \dfrac{1}{3}\sqrt{54}\right)$

13 $\dfrac{n}{m}\sqrt{\dfrac{n}{2m^3}} \times \left(-\dfrac{1}{m}\sqrt{\dfrac{n^3}{m^3}}\right) \div \sqrt{\dfrac{n}{2m^3}}$

14 $-3\sqrt{\dfrac{3m^2 - 3n^2}{2a^2}} \div \left(\dfrac{3}{2}\sqrt{\dfrac{m+n}{a^2}}\right) \times \sqrt{\dfrac{a^2}{m-n}}$

15 Given that x and y are real numbers and $y = \dfrac{\sqrt{x^2 - 4} + \sqrt{4 - x^2} + 1}{x + 2}$, find the value of $\sqrt{x + y} \times \sqrt{x - y}$.

16 Given that $x = \sqrt{2} - 1$, find the value of $x^3 + 2x^2 + x + 2$.

17 Given that $\sqrt{\dfrac{9 - x}{x - 6}} = \dfrac{\sqrt{9 - x}}{\sqrt{x - 6}}$ and x is an even number, find the value of $(1 + x)\sqrt{\dfrac{x^2 - 5x + 4}{x^2 - 1}}$.

5.11 Rationalising denominators

Learning objectives

Know how to rationalise denominators and apply this to solve problems.

A. Fill in the blanks

Rationalise the denominator in each expression.

1 $\dfrac{1}{\sqrt{8}} = $ _____

2 $\dfrac{1}{\sqrt{2}-1} = $ _____

3 $\dfrac{2}{1+\sqrt{3}} = $ _____

4 $\dfrac{1}{\sqrt{3}-2} = $ _____

5 $\dfrac{1}{\sqrt{18m}} = $ _____

6 $\dfrac{1}{\sqrt{x+y}} = $ _____

7 $\dfrac{x-y}{\sqrt{x}+\sqrt{y}} = $ _____

8 $\dfrac{2a}{\sqrt{7a}-\sqrt{5a}} = $ _____

B. Short-answer questions

Rationalise the denominator in each expression.

9 $\dfrac{2-\sqrt{3}}{2+\sqrt{3}}$

10 $\sqrt{\dfrac{x+y}{x-y}} \ (x > y > 0)$

11 $\dfrac{m+n-2\sqrt{mn}}{\sqrt{m}-\sqrt{n}}$

12 $\dfrac{\sqrt{x+1}-\sqrt{x-1}}{\sqrt{x+1}+\sqrt{x-1}}$

13 Simplify: $\dfrac{1}{\sqrt{3}-\sqrt{2}}+\dfrac{1}{\sqrt{3}+\sqrt{2}}$

14 Compare the values of $\dfrac{1}{\sqrt{7}-\sqrt{5}}$ and $\dfrac{1}{\sqrt{8}-\sqrt{6}}$.

15 Given that the sum of $(a+\sqrt{2})^2$ and $\sqrt{(b+1)^2}$ is zero, find the value of $\dfrac{1}{b-a}$.

16 Given that $x=\dfrac{1}{2-\sqrt{3}}$, find the value of the algebraic expression x^2-4x+2.

17 Given that $x=\sqrt{2008-2a}+\sqrt{a-1004}+5$, where a is a real number, simplify $\dfrac{\sqrt{x+1}-\sqrt{x}}{\sqrt{x+1}+\sqrt{x}}+\dfrac{\sqrt{x+1}+\sqrt{x}}{\sqrt{x+1}-\sqrt{x}}$, and then find its value.

18 Calculate: $(2\sqrt{5}+1)\left(\dfrac{1}{1+\sqrt{2}}+\dfrac{1}{\sqrt{2}+\sqrt{3}}+\dfrac{1}{\sqrt{3}+\sqrt{4}}+\cdots+\dfrac{1}{\sqrt{99}+\sqrt{100}}\right)$.

5.12　Mixed operations with quadratic surds

Learning objectives

Know how to add, subtract, multiply and divide expressions involving quadratic surds.

A. Fill in the blanks

1　$\sqrt{50} + (-\sqrt{18}) = $ _____

2　$-5x\sqrt{\dfrac{a}{x}} + \sqrt{4ax} = $ _____　$(x > 0)$

3　Given that $a = \sqrt{7} + 2$ and $b = \sqrt{7} - 2$, then $a + b = $ _____

4　$(3 - \sqrt{2})(2 + \sqrt{3}) = $ _____

5　$\dfrac{1 - \sqrt{2}}{2} \cdot \dfrac{1 + \sqrt{2}}{2} = $ _____

6　$(\sqrt{2} - \sqrt{12})(\sqrt{18} + \sqrt{48}) = $ _____

7　$(\sqrt{18} - 2\sqrt{2}) \cdot \sqrt{\dfrac{1}{12}} = $ _____

8　$(\sqrt{12} - 2\sqrt{18})^2 = $ _____

B. Short-answer questions

Simplify the expressions in Questions 9–14.

9　$\left(5\sqrt{\dfrac{1}{2}} - 6\sqrt{\dfrac{3}{2}}\right)\left(\dfrac{1}{4}\sqrt{8} - \sqrt{\dfrac{2}{3}}\right)$

10　$\left(\dfrac{1}{2}\sqrt{3} + \sqrt{8}\right)\left(\sqrt{8} - \dfrac{1}{2}\sqrt{3}\right)$

11 $(10\sqrt{48} - 6\sqrt{27} + 4\sqrt{12}) \div \sqrt{6}$

12 $\sqrt{2} \times \left(\sqrt{2} + \dfrac{1}{\sqrt{2}}\right) - \dfrac{\sqrt{18} - \sqrt{8}}{\sqrt{2}}$

13 $(1 + \sqrt{2})^{2016} \times (1 - \sqrt{2})^{2017}$

14 $(\sqrt{a} + \sqrt{b})^2 - (\sqrt{a} - \sqrt{b})^2$

15 Given that $x = \sqrt{3} + \sqrt{2}$ and $y = \sqrt{3} - \sqrt{2}$, find the value of:
(a) $x^2 - xy + y^2$
(b) $x^3 y + xy^3$.

16 Given that $x = \sqrt{5} - 2$, find the value of $(9 + 4\sqrt{5})x^2 - (\sqrt{5} + 2)x + 4$.

17 Given that the integer part of $\dfrac{\sqrt{3} + 1}{\sqrt{3} - 1}$ is a and the decimal part is b, find the value of $a^2 + ab + b^2$.

Unit test 5

A. Multiple choice questions

1 If $\sqrt{x-2}$ is meaningful, then x must satisfy the condition ( ).

A. $x > 2$

B. $x \geqslant 2$

C. $x < 2$

D. $x \leqslant 2$

2 Of these pairs of quadratic surds, the pair of like quadratic surds is ( ).

A. $\sqrt{\dfrac{1}{xy}}$ and $\sqrt{\dfrac{1}{2xy}}$

B. $\sqrt{8ab^3}$ and $2\sqrt{ab}$

C. $\sqrt{20}$ and $-\sqrt{\dfrac{1}{5}}$

D. $\sqrt{a}$ and $\sqrt{ab}$

3 Of these calcuations, the incorrect one is ( ).

A. $\sqrt{14} \times \sqrt{7} = 7\sqrt{2}$

B. $\sqrt{60} \div \sqrt{5} = 2\sqrt{3}$

C. $\sqrt{9a} + \sqrt{25a} = 8\sqrt{a}$

D. $3\sqrt{2} - \sqrt{2} = 3$

4 Of these equations, the incorrect one is ( ).

A. $(a^2 b)^{\frac{1}{n}} = a^{\frac{2}{n}} b^{\frac{1}{n}}$

B. $\left(\dfrac{a^2}{b}\right)^{\frac{1}{n}} = a^{\frac{2}{n}} b^{-\frac{1}{n}}$

C. $\sqrt[n]{a^2 + b} = (a^2 + b)^{\frac{1}{n}}$

D. $\sqrt[n]{a^2 - b} = a^{\frac{2}{n}} - b^{\frac{1}{n}}$

5 Given that $\sqrt{24n}$ is an integer, then the least possible value of the positive integer n is ( ).

A. 2

B. 6

C. 8

D. 9

6 Given that $a = 2 + \sqrt{3}$ and $b = \dfrac{1}{2 - \sqrt{3}}$, then ( ).

A. $a = b$

B. $a > b$

C. $a < b$

D. $a = \dfrac{1}{b}$

7 The diagram shows a cubiod box with length 4 units, width 4 units and height 6 units. If an ant crawls on the surface of the box from point A to point B, then the length of the shortest route is (　　).

A. 9 units

B. 10 units

C. $4\sqrt{2}$ units

D. $2\sqrt{17}$ units

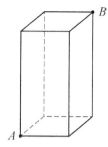

Diagram for question 7

B. Fill in the blanks

8 When a is _____, $\dfrac{\sqrt{a^2}}{a} = -1$.

9 The condition for the equation $\sqrt{a^2 - 9} = \sqrt{a + 3} \times \sqrt{a - 3}$ to be true is _____.

10 Calculate: $\left(2\dfrac{1}{4}\right)^{-\frac{1}{2}} + \left(\dfrac{1}{\sqrt{3}}\right)^{-2} = $ _____.

11 Given that $0 < a < b$, calculate: $\sqrt{a + b + 2\sqrt{ab}} \times \sqrt{a + b - 2\sqrt{ab}} = $ _____.

12 Given that $m < 0$, simplify: $2n\sqrt{\dfrac{m}{n}} = $ _____.

13 Calculate: $\sqrt{18a} - \sqrt{\dfrac{1}{8}a} + 4\sqrt{0.5a} = $ _____.

14 Calculate: $(3\sqrt{2} - 2\sqrt{3})(3\sqrt{2} + 2\sqrt{3}) = $ _____.

15 Calculate: $(3\sqrt{6} - \sqrt{15})^2 = $ _____.

16 Calculate: $\sqrt{3 + 2\sqrt{2}} \times \sqrt{3 - 2\sqrt{2}} = $ _____.

17 Given that $a + b = 5$ and $ab = 4$, then $\dfrac{\sqrt{a} - \sqrt{b}}{\sqrt{a} + \sqrt{b}} = $ _____.

18 Given that $a < 1$ and $a \neq 0$, simplify: $\dfrac{\sqrt{a^2 - 2a + 1}}{a^2 - a} = $ _____.

19 The diagram shows the position of a real number a on the number line. Simplify $\sqrt{(a-1)^2} + \sqrt{(a-2)^2} = $ _____ .

$$\overrightarrow{\begin{array}{ccccccc} | & | & | & | & \overset{a}{\underset{\bullet}{|}} & | \\ -1 & 0 & 1 & & 2 & \end{array}}$$

Diagram for question 19

20 The lengths of the three sides of a triangle are $\sqrt{8}$ cm, $\sqrt{12}$ cm and $\sqrt{18}$ cm. Its perimeter is _____ cm.

C. Questions that require solutions

Complete the calculations in Questions 21–24.

21 $\left(\dfrac{2-\sqrt{6}}{3}\right)^2 \times (5-2\sqrt{6})^{-1}$

22 $\dfrac{2}{3}\sqrt{9x} + 6\sqrt{\dfrac{x}{4}} - 2x\sqrt{\dfrac{1}{x}}$

23 $(\sqrt{2}+\sqrt{3}+\sqrt{6})(\sqrt{2}-\sqrt{3}+\sqrt{6})$

24 $\dfrac{2}{b}\sqrt{ab} \times \left(-\dfrac{3}{2}\sqrt{a^3b}\right) \div \dfrac{1}{3}\sqrt{\dfrac{b}{a}}$ $(a>0)$

25 Find the value of the algebraic expression $a^2 + ab - b^2$, when $a = \sqrt{2}+1$ and $b = \sqrt{3}-1$.

26 Given that $x + y = 3$ and $xy = 6$, find the value of $\sqrt{\dfrac{x}{y}} + \sqrt{\dfrac{y}{x}}$.

㉗ Given that $a + \dfrac{1}{a} = 1 + \sqrt{10}$, find the value of $a^2 + \dfrac{1}{a^2}$.

28 Given that $x^2 - x - 2 = 0$, find the value of $\dfrac{x^2 - x + 2\sqrt{3}}{(x^2 - x)^2 - 1 + \sqrt{3}}$.

㉙ Look at these equations.

① $\sqrt{9 \times 9 + 19} = 10$ ② $\sqrt{99 \times 99 + 199} = 100$

③ $\sqrt{999 \times 999 + 1999} = 1000$

(a) Based on the pattern shown in these three equations, write the fourth equation and then use calculation to verify that the equation is true.

(b) Follow the pattern you have noticed in the equations to write the nth equation.

30* There is a type of question in algebra, that involves simplifying expressions of the form $\sqrt{a \pm 2\sqrt{b}}$. If you can find two numbers m and n such that $m^2 + n^2 = a$ and $mn = \sqrt{b}$, then you can convert $a \pm 2\sqrt{b}$ to $m^2 + n^2 \pm 2mn = (m \pm n)^2$. Then, by extracting the root, you can simplify $\sqrt{a \pm 2\sqrt{b}}$.

> **Example**: Simplify $\sqrt{3 \pm 2\sqrt{2}}$.
> Since $3 + 2\sqrt{2} = 1 + 2 + 2\sqrt{2} = 1^2 + (\sqrt{2})^2 + 2\sqrt{2} = (1 + \sqrt{2})^2$, then:
> $\sqrt{3 + 2\sqrt{2}} = \sqrt{(1 + \sqrt{2})^2} = 1 + \sqrt{2}$.

Follow the example and simplify each expression.

(a) $\sqrt{7 + 4\sqrt{3}}$ (b) $\sqrt{13 - 2\sqrt{42}}$

∗ Challenging questions (optional).

Chapter 6　Quadratic equations

6.1　Concepts of quadratic equations

Learning objective

Understand the concepts of quadratic equations in one variable; know all the relevant terms.

A. Multiple choice questions

1 Of these equations, (　　) is a quadratic equation in x.

　A. $3(x + 1)^2 = 2(x + 1)$

　B. $\dfrac{1}{x^2} + \dfrac{1}{x} - 2 = 0$

　C. $ax^2 + bx + c = 0$

　D. $x^2 + 2x = x^2 - 1$

2 Of these equations, (　　) are quadratic equations in one variable.

　① $2x^2 = -3x$　　② $3x^2(x - 3) = x$　　③ $(3x - 2)\sqrt{3}(y^2 - 1) = \sqrt{15}$

　④ $\dfrac{2}{x^2} - 1 = 0$　　⑤ $\dfrac{y^2}{4} = 7$　　⑥ $x^2 = 9$

　A. ①, ③ and ⑤

　B. ①, ⑤ and ⑥

　C. ①, ③ and ④

　D. ②, ④ and ⑤

B. Fill in the blanks

3 When $5x^2 = 6x - 8$ is rearranged into the standard form of a quadratic equation in one variable, the coefficient of its quadratic term is _____, the coefficient of its linear term is _____ and its constant term is _____.

4 Given that 3 is a solution to the equation $\dfrac{4}{3}x^2 - 2a + 1 = 0$ in x, the value of $2a$ is _____.

5 Given an equation $ax^2 + bx + c = 0$ ($a \neq 0$), in which $a - b + c = 0$, then the equation must have a root that is _____.

6 Given that the equation $mx^2 + 3x - 4 = 0$ is a quadratic equation in one variable x, then the set of values that m can take is _____.

7 When k _____, the equation $(k - 3)x^2 + 2x - 1 = 0$ in x is a quadratic equation in one variable.

8 When the quadratic equation in one variable, $x(x + 3) = 2x - 5$, is rearranged into standard quadratic form, it is _____.

9 The coefficient of the quadratic term of equation $3x^2 = x$ is _____, the coefficient of its linear term is _____ and the constant term is _____.

10 Given that $(a^2 - 1)x^3 - (a + 1)x^2 + 4 = 0$ is a quadratic equation in one variable x, then the value(s) that a can take is/are _____.

C. Questions that require solutions

11 Which of the numbers in brackets after each equation is/are the solution(s) to the equation?

(a) $3x^2 - 2x - 1 = 0$, $\left(\sqrt{2}, 1, -\dfrac{1}{3}\right)$ (b) $2x^2 - 3x + 1 = 0$, $\left(\dfrac{1}{2}, 1, 2\right)$

12 Consider the equation $kx^2 - k(x + 2) = x(2x + 3) + 1$ in x.

(a) For what value(s) of k is this equation a quadratic equation in one variable?

(b) For what value(s) of k is this equation a linear equation in one variable?

(c) Is -1 a root of the equation? Why?

13 Given the equation $(m - 2)x^{-m} + 2x + 4 = 2m - 1$ in x, for what value(s) of m is it a quadratic equation in one variable and a linear equation in one variable, respectively?

14 Consider the equation $x^2 - mx(2x - m - 1) = 0$.

(a) For what value(s) of m is it a quadratic equation in one variable?

(b) If one root of the equation is $x = 0$, find the value of m.

6.2 Solving quadratic equations (1): by taking square roots

Learning objective

Know how to solve quadratic equations by taking square roots.

A. Multiple choice questions

1. If the value of the algebraic expression $3x^2 - 6$ is 21, then the value of x must be ( ).

   A. 3 B. ± 3

   C. -3 D. $\pm\sqrt{3}$

2. The solutions to the equation $(x + 2)^2 = 4$ are ( ).

   A. $x = 4$ and -4 B. $x = 4$ and 0

   C. $x = -4$ and 0 D. $x = 0$ and 2

B. Fill in the blanks

3. The solutions to the equation $x^2 - 4 = 0$ are _____.

4. Given that $4x^2 - 8 = 0$, then $x =$ _____.

5. Given the equation $2x^2 + 8 = 0$ in x, then x _____.

6. If the equation $(x - a)^2 + b = 0$ in x has real number solutions, then the set of values that b can take is _____.

7. If the equation $ax^2 - b = 0$ in x has real solutions, then a and b satisfy the conditions that _____.

8. If both roots of the equation $(3x - c)^2 - 60 = 0$ are positive numbers, and c is an integer, then the smallest value of c is _____.

C. Questions that require solutions

9 Solve these quadratic equations by taking square roots.

(a) $x^2 - 49 = 0$

(b) $2x^2 - 18 = 0$

(c) $3y^2 - 5 = 0$

(d) $(x - \sqrt{5})^2 = 4$

(e) $(2x + 1)^2 - 32 = 0$

(f) $\sqrt{3}(x + 6)^2 - 27\sqrt{3} = 0$

(g) $(2x + 1)^2 = (3x - 2)^2$

(h) $(2x - 1)^2 = (x + 1)^2$

(i) $(x - a)^2 = a^2 + 2ab + b^2$

(j) $(x - m)^2 = n$

10 Given that the equation $(2x - 1)^2 = a$ has two unequal real roots, the equation $(2x - 1)^2 = b$ has two equal real roots and the equation $(2x - 1)^2 = c$ has no real roots, compare the values of a, b and c.

6.3 Solving quadratic equations (2): by factorising

 Learning objective

Know how to solve quadratic equations by factorisation.

 A. Multiple choice questions

① The root(s) of the equation $2x(x - 3) = 5(x - 3)$ is/are (   ).

   A. $x = \dfrac{5}{2}$ B. $x = 3$

   C. $x_1 = \dfrac{5}{2}$ and $x_2 = 3$ D. $x = \dfrac{2}{5}$

② The solution(s) to the equation $x(x - 1) = 2$ is/are (   ).

   A. $x = -1$ B. $x = -2$

   C. $x_1 = 1$ and $x_2 = -2$ D. $x_1 = -1$ and $x_2 = 2$

 B. Fill in the blanks

③ The result of factorising $x^2 - 5x$ is _____.

④ The result of factorising the quadratic expression $x^2 + 20x + 96$ is _____. When $x^2 + 20x + 96 = 0$, its two roots are _____.

⑤ The roots of the equation $x^2 + 6x - 7 = 0$ are _____.

⑥ The roots of the equation $(2x - 1)^2 = 2x - 1$ are _____.

⑦ If the quadratic equation $(m + 3)x^2 + x - m^2 - 5m - 6 = 0$ in one variable x has one root equal to 0, then $m = $ _____.

⑧ The solutions to the equation $x^2 = 2\sqrt{x^2}$ is _____.

C. Questions that require solutions

9 Write the quadratic equations in one variable that satisfy the given conditions.

(a) The two roots are -3 and 6 and the coefficient of the quadratic term is 1.

(b) The two roots are $\sqrt{5} - 4$ and $\sqrt{5} + 4$ and the coefficient of the linear term is 2.

10 Solve these equations by factorising.

(a) $5x^2 = 4x$

(b) $4x^2 - 25 = 0$

(c) $x - 2 = x(x - 2)$

(d) $(x + 1)^2 - 25 = 0$

(e) $y(y + 5) = 24$

(f) $(x + 4)^2 - (2x - 1)^2 = 0$

(g) $(3 - y)^2 + y^2 = 9$

(h) $2x^2 + \sqrt{3}x + 2x = 3 + \sqrt{3}$

11 Given that the lengths of two sides of an isosceles triangle are the two roots of the equation $x^2 - 6x + 8 = 0$, find the perimeter of the triangle.

12 Given that the roots of the equation $(a - 1)x^2 - (a^2 + 1)x + a^2 + a = 0$ are positive integers, find the integer value of a.

6.4 Solving quadratic equations (3): by completing the square

Learning objective

Know how to solve quadratic equations by completing the square.

A. Multiple choice questions

1. Completing the square, the equation $x^2 + 3 = 4x$ becomes (　　).

   A. $(x - 2)^2 = 7$　　　　　　　　　B. $(x + 2)^2 = 21$

   C. $(x - 2)^2 = 1$　　　　　　　　　D. $(x + 2)^2 = 2$

2. Solving the equation $x^2 + 4x = 10$ by completing the square gives the root(s) of the equation as (　　).

   A. $2 \pm \sqrt{10}$　　　　B. $-2 \pm \sqrt{14}$　　　　C. $-2 + \sqrt{10}$　　　　D. $2 - \sqrt{10}$

B. Fill in the blanks

3. $x^2 - 8x +$ _____ $= (x -$ _____$)^2$.

4. $x^2 - 6mx +$ _____ $= (x -$ _____$)^2$.

5. $x^2 - 2(a + b)x +$ _____ $= (x -$ _____$)^2$.

6. Completing the square in the quadratic expression $x^2 - 2x - 2$ * gives the result _____.

7. If $x^2 + 6x + m^2$ is a perfect square, then the value of m is _____.

8. Completing the square on the left side of the equation $3x^2 + \sqrt{2}x - 6 = 0$ gives the new equation _____.

---

*　Such a quadratic expression with three terms is also known as a quadratic trinomial.

9 Given that $x^2 + y^2 + 4x - 6y + 13 = 0$, then $x + y = $ _____.

10 No matter what real values x and y take, the value of the polynomial $x^2 + y^2 - 2x - 4y + 16$ is always a _____ number.

C. Questions that require solutions

11 Solve these equations by completing the square.

(a) $x^2 - 2x - 5 = 0$

(b) $x^2 - 4x = 9996$

(c) $x^2 - 0.6x - 0.16 = 0$

(d) $2x^2 + 3x - 7 = 0$

(e) $x^2 + \dfrac{1}{6}x - \dfrac{1}{3} = 0$

(f) $\dfrac{2}{3}y^2 + \dfrac{1}{3}y - 2 = 0$

(g) $0.25x^2 + x - 1 = 0$

(h) $3x^2 - 2 = 4x$

(i) $x^2 - 5ax - 6a^2 = 0$

(j) $y^2 + 2(\sqrt{3} + 1)y + 2\sqrt{3} = 0$

12 Given that $x^2 + 6x + y^2 - 8y + 25 = 0$, find the value of $\dfrac{x - 2y}{x^2 + y^2}$.

6.5 Solving quadratic equations (4): by using the quadratic formula

Learning objective

Know how to solve quadratic equations by using the quadratic formula.

A. Multiple choice questions

1. Using the quadratic formula to solve the equation $3x^2 + 4 = 12x$, the correct solutions are (　　).

    A. $x = \dfrac{12 \pm \sqrt{12^2 - 4 \times 3 \times 4}}{2 \times 3}$

    B. $x = \dfrac{-12 \pm \sqrt{12^2 - 4 \times 3 \times 4}}{2 \times 3}$

    C. $x = \dfrac{12 \pm \sqrt{12^2 + 4 \times 3 \times 4}}{2 \times 3}$

    D. $x = \dfrac{-(-12) \pm \sqrt{-(-12)^2 - 4 \times 3 \times 4}}{2 \times 3}$

2. The solutions to the equation $x^2 + 3x = 14$ are (　　).

    A. $x = \dfrac{3 \pm \sqrt{65}}{2}$ 　　　　　　B. $x = \dfrac{-3 \pm \sqrt{65}}{2}$

    C. $x = \dfrac{3 \pm \sqrt{23}}{2}$ 　　　　　　D. $x = \dfrac{-3 \pm \sqrt{23}}{2}$

B. Fill in the blanks

3. The solutions to the quadratic equation in one variable $ax^2 + bx + c = 0$ ($b^2 - 4ac \geqslant 0$) are _____ .

4. For the equation $x^2 + x - 2 = 0$, the value of $b^2 - 4ac$ is _____ .

⑤ For the equation $3x^2 + 2\sqrt{3}x = 2$, the value of $b^2 - 4ac$ is _____.

6 If the values of the algebraic expressions $4x^2 - 2x - 5$ and $2x^2 + 1$ sum to zero, then the value of x is _____.

C. Questions that require solutions

⑦ Solve the equations by using the quadratic formula.

(a) $x^2 - 3x - 4 = 0$

(b) $x^2 = 2(x - 1)$

(c) $x^2 - 4\sqrt{3}x + 10 = 0$

(d) $-5x^2 - 3x + 2 = 0$

(e) $5x^2 - 2x - 7 = 0$

(f) $\frac{1}{2}(x - 1) = 2x^2$

(g) $x^2 + 2\sqrt{3}x + 3 = 0$

(h) $(2x - 7)^2 - (2x + 5)^2 = 12x^2$

(i) $2x^2 + 2\sqrt{7}x + 1 = 0$

(j) $(x + 1)(x - 1) = 2\sqrt{2}x$

8 Use the quadratic formula to solve the following equations in x.

(a) $x^2 - x = a^2$

(b) $20m^2x^2 + 11mnx - 3n^2 = 0$ $(m \neq 0)$

6.6　Solving quadratic equations (5): by suitable methods

Learning objective

Identify and use suitable methods to solve quadratic equations.

A. Multiple choice questions

1. Solving the equation $4y^2 = 12y + 3$ gives (　　).

   A. $y = \dfrac{-3 \pm \sqrt{6}}{2}$　　　B. $y = \dfrac{3 \pm \sqrt{6}}{2}$　　　C. $y = \dfrac{3 \pm 2\sqrt{3}}{2}$　　　D. $y = \dfrac{-3 \pm 2\sqrt{3}}{2}$

2. Let a be the larger root of the quadratic equation $x^2 + 5x = 0$ and b be the smaller root of the equation $x^2 - 3x + 2 = 0$, then the value of $a + b$ is (　　).

   A. -4　　　　　　B. -3　　　　　　C. 1　　　　　　D. 2

B. Fill in the blanks

3. Using the method of _____ to solve the equation $(x - 1)^2 = 4$ gives the solutions $x_1 = $ _____ and $x_2 = $ _____.

4. Using the method of _____ to solve the equation $(x - 2)^2 - 2 + x = 0$ gives the solutions $x_1 = $ _____ and $x_2 = $ _____.

5. Using the method of _____ to solve the equation $x^2 - 2x = 399$ gives the solutions $x_1 = $ _____ and $x_2 = $ _____.

6. If the value of the algebraic expression $2x^2 - 3x - 5$ is equal to the value of the algebraic expression $4 - 6x$, then $x = $ _____.

7. Given that -2 is one of the roots of the quadratic equation in one variable x, $x^2 + 2kx + k^2 = 0$, then $k = $ _____.

8. The two roots of the quadratic equation in one variable x, $x^2 + 2x + c = 0$ ($c \leqslant 1$), are _____.

C. Questions that require solutions

9 Use suitable methods to solve these quadratic equations in one variable.

(a) $(3x + 1)^2 = 4$

(b) $x^2 - 4x - 60 = 0$

(c) $5(x - 3)^2 + x = 3$

(d) $5x^2 - 2x - 7 = 0$

(e) $\frac{1}{2}x - 1 = 2x^2$

(f) $3x^2 + 2\sqrt{3}x + 1 = 0$

(g) $3(x - 2)^2 = x^2 - 2x$

(h) $(x + 3)^2 = (2x - 5)^2$

(i) $\sqrt{2}y^2 + 4\sqrt{3}y = 2\sqrt{2}$

(j) $(3 - 2\sqrt{2})x^2 + 2(\sqrt{2} - 1)x - 3 = 0$

(k) $ax^2 - (bc + ca + ab)x + b^2c + bc^2 = 0 \ (a \neq 0)$

(l) $mx^2 + (4m + 1)x + 4m + 2 = 0$

10 Given that $y = 2x^2 + 7x - 1$, for what value of x is the value of y equal to the value of $4x + 1$? For what value of x do the value of y and the value of $x^2 - 19$ sum to zero?

11 Given that a is a root of the equation $x^2 - x - 1 = 0$, find the value of the algebraic expression $a^3 - 2a + 3$.

12 Given that $x^2 = 1 - x$ $(x > 0)$, find the value of $x + \dfrac{1}{x}$.

6.7 Discriminant of a quadratic equation (1)

Learning objective

Understand the meaning of the discriminant of a quadratic equation and its relationship to the nature of the roots.

A. Multiple choice questions

1 Of these quadratic equations in one variable x, (   ) has two distinct real roots.

A. $x^2 + 1 = 0$

B. $x^2 + 2x + 1 = 0$

C. $x^2 + 2x + 3 = 0$

D. $x^2 + 2x - 3 = 0$

2 Of these quadratic equations in one variable x, (   ) has two distinct real roots.

A. $x^2 + 2 = 0$

B. $x^2 + x - 1 = 0$

C. $x^2 + x + 3 = 0$

D. $4x^2 - 4x + 1 = 0$

B. Fill in the blanks

3 In the quadratic equation in one variable x, $ax^2 + bx + c = 0$ ($a \neq 0$), the discriminant $\Delta = $ _____. When Δ is _____, the equation has two distinct real roots. When Δ is _____, the equation has two equal real roots. When $\Delta < 0$, the equation has _____ real roots. When $\Delta \geq 0$, the two roots of the equation are $x_1 = $ _____ and $x_2 = $ _____.

4 The value of the discriminant of the equation $2x^2 + 4x - 1 = 0$ is _____.

5 For the quadratic equation $\frac{2}{3}x - x^2 - \frac{1}{3} = 0$, the discriminant Δ _____ 0, hence it has _____ real roots.

6 The value of the discriminant of the equation $x^2 - x = \frac{1}{2}$ is _____. The equation has _____ root(s).

7 Given that the quadratic equation in one variable x, $x^2 + 4x + a = 0$, has two equal real roots, then $a = $ _____.

8 Given that the algebraic expression $x^2 - 2(m + 1)x + m^2 + 5$ is a perfect square, then $m = $ _____.

C. Questions that require solutions

9 Without solving the equations, determine the nature of their roots.

(a) $2x^2 + 3x = 4$

(b) $3x^2 = 2(2x - 1)$

(c) $7x^2 + 1 = 2\sqrt{7}x$

(d) $4p(p - 1) - 3 = 0$

(e) $\sqrt{3}x^2 + x = \sqrt{2}$

(f) $(2x - 1)^2 + x(x + 2) = 0$

10 Without solving these equations in x, determine the nature of their roots.

(a) $ax^2 - bx = 0 \ (a \neq 0)$

(b) $x^2 - 2\sqrt{3}x + 3k^2 + 7 = 0$

(c) $x^2 - mx + \dfrac{1}{2}m^2 + m + \dfrac{3}{2} = 0$

(d) $(a^2 + b^2)x^2 - 2b(a + c)x + (b^2 + c^2) = 0 \ (a \neq 0, \ b^2 = ac)$

11 Given that $x^2 - 2mx + 4(m - 1) = 0$ is a quadratic equation in one variable x, does it have real roots? Why?

12 Given that the equation $x^2 - 2x - m = 0$ has no real roots and m is a real number, determine whether the equation $x^2 + 2mx + m(m + 1) = 0$ has real roots.

6.8 Discriminant of a quadratic equation (2)

Learning objective

Use the discriminant of a quadratic equation to determine the roots of the equation.

A. Multiple choice questions

1 Given that the quadratic equation in one variable x, $x^2 - 2x + a = 0$, has real root(s), the set of values that the real number a can take is (　　).

A. $a \leqslant 1$ B. $a < 1$ C. $a \leqslant -1$ D. $a \geqslant 1$

2 Given that the quadratic equation in one variable x, $2x^2 - 2\sqrt{2}x + m = 0$, has two real roots, the result of simplifying $\sqrt{(m-1)^2}$ is (　　).

A. $m - 1$ B. $1 - m$ C. $\pm (m - 1)$ D. $m + 1$

B. Fill in the blanks

3 The value of the discriminant of the quadratic equation in one variable x, $2x^2 - (2m + 1)x + m = 0$, is 9. Then $m =$ _____.

4 The equation in x, $mx^2 - 2x + 1 = 0$, has two real roots. The set of values that m can take is _____.

5 Given that the quadratic equation in one variable x, $x^2 - x - m = 0$, has two distinct real roots, the set of possible values of m is _____.

6 If the quadratic equation in one variable x, $2x^2 - (4k + 1)x + 2k^2 - 1 = 0$, has no real roots, then the set of values that k can take is _____.

7 If the quadratic equation in one variable x, $kx^2 - 3x + 2 = 0$, has real roots, then the set of values that k can take is _____.

8 Given the quadratic equation in one variable x, $x^2 + (3 - m)x + \dfrac{m^2}{4} = 0$, has two real roots, the maximum integer value of m is _____.

C. Questions that require solutions

9 For what values of k does the quadratic equation in one variable x, $x^2 - 4x + k - 5 = 0$, satisfy these conditions?

(a) It has two distinct real roots.

(b) It has two equal real roots.

(c) It has no real roots.

10 Prove that the equation $(m^2 + 1)x^2 - 2mx + (m^2 + 4) = 0$ has no real roots.

11 Given that the equation in x, $(k^2 - 1)x^2 + 2(k + 1)x + 1 = 0$, has real roots, find the set of values that k can take.

12 The equation in x, $x^2 - \sqrt{2k + 4}x + k = 0$, has two distinct real roots.

(a) Find the set of value that k can take.

(b) Simplify $\sqrt{k^2 + 4k + 4} + \sqrt{k^2 - 4k + 4}$.

13 a, b and c are the lengths of the three sides of $\triangle ABC$ and the equation in x, $(b + c)x^2 + \sqrt{2}(a - c)x - \dfrac{3}{4}(a - c) = 0$, has two equal real roots. Prove that $\triangle ABC$ is an isosceles triangle.

6.9 Applications of quadratic equations (1): factorising quadratic expressions

 Learning objective

Factorise quadratic expressions by solving the related quadratic equation.

 A. Multiple choice questions

1 If the two roots of the quadratic equation in one variable x, $x^2 + px + q = 0$, are 3 and 4, then the quadratic expression $x^2 + px + q$ can be factorised as (   ).

A. $(x+3)(x-4)$ 　　　　　　　　B. $(x-3)(x+4)$

C. $(x-3)(x-4)$ 　　　　　　　　D. $(x+3)(x+4)$

2 The result of factorising $2x^2 - 8xy + 5y^2$ is (   ).

A. $2\left(x - \dfrac{4 + \sqrt{6}}{2}\right)\left(x - \dfrac{4 - \sqrt{6}}{2}\right)$ 　　　　B. $\left(x - \dfrac{4 + \sqrt{6}}{2}y\right)\left(x - \dfrac{4 - \sqrt{6}}{2}y\right)$

C. $2\left(x - \dfrac{4 + \sqrt{6}}{2}y\right)\left(x - \dfrac{4 - \sqrt{6}}{2}y\right)$ 　　　　D. $(2x - 4y - \sqrt{6}y)(2x - 4y + \sqrt{6}y)$

 B. Fill in the blanks

3 Given that the equation $3x^2 + 4x - 1 = 0$ has two roots $x_1 = \dfrac{-2 + \sqrt{7}}{3}$ and $x_2 = \dfrac{-2 - \sqrt{7}}{3}$, the result of factorising the quadratic expression $3x^2 + 4x - 1$ is

_____.

4 Solving the equation $3x^2 - 4xy - 4y^2 = 0$, the roots are $x_1 =$ _____ and $x_2 =$ _____. Therefore, the result of factorising $3x^2 - 4xy - 4y^2$ is _____.

5 Factorise the quadratic expression: $-2x^2 - 3x + 6 =$ _____.

6 Given that the polynomial $x^2 + kx + 5(k - 5)$ in x is a perfect square, then the value of k is _____.

7 Given that the quadratic expression $2x^2 - 3x + m + 1$ can be factorised, then the set of values that m can take is _____.

8 When the polynomial $3kx^2 + (6k - 1)x + 3k + 1 (k \neq 0)$ can be factorised, then the set of values that k can take is _____.

C. Questions that require solutions

9 Factorise these equations.

(a) $x^2 - 5x + 6$ (b) $4x^2 - 5$ (c) $4x^2 + 8x - 1$ (d) $3t^2 - 4t - 1$

(e) $2y^2 - 4y + 1$ (f) $2x^2 - 8xy + 5y^2$ (g) $3x^2y^2 - 5xy - 1$ (h) $-4x^2 - 8xy + y^2$

(i) $-y^2x - 4x^2y + 2x^3$ (j) $(x^2 + x)^2 - 2x(x + 1) - 3$

10 In an isosceles triangle, the length of each of its two equal sides is a, and the length of its other side is b. Prove that the quadratic expression in x, $x^2 - 4ax + b^2$, can be factorised.

11 Given that the result of factorising the quadratic expression in x, $4x^2 - kx + 1$, is $(2x - \sqrt{2} + 1)(tx + m)$, find the values of the real numbers k, m and t.

6.10 Applications of quadratic equations (2): practical applications

Learning objective

Use quadratic equations to solve practical problems.

A. Multiple choice questions

1 The total output of a factory in January this year was 500 tonnes and the total output in March was 720 tonnes. Taking the average monthly growth rate as x, then we can establish the equation (   ).

A. $500(1 + 2x) = 720$

B. $500(1 + x)^2 = 720$

C. $500(1 + x^2) = 720$

D. $720(1 + x)^2 = 500$

2 The side length of a square is 4 cm greater than half the side length of another, larger square. The area of the larger square is 32 cm² less than twice the area of the smaller square. Then the side lengths of the two squares are (   ) respectively.

A. 18 cm and 10 cm

B. 17 cm and 11 cm

C. 16 cm and 12 cm

D. 15 cm and 13 cm

B. Fill in the blanks

3 A 13-metre length of fencing is used to enclose a rectangular piece of land with area 20 m² that is bounded on one side by a wall. The lengths of the two sides of the fence are _____ metres, respectively.

4 Each member in a group sends a picture to all the other members. As a result, a total of 90 pictures were exchanged. There are _____ members in the group.

5 Joshua deposited £1000 into a one-year fixed deposit account with a bank. When the year was up, he withdrew £200 and deposited the remaining £800, together with the interest earned, for another year with the same terms and conditions. Let the annual interest rate be x. Joshua expects to get £892.50 in total, including the principal and the interest, at the end of this period. This can be described by the equation:

_____.

6 The value of a two-digit number equals three times the square of the ones-digit. The tens-digit is 2 greater than the digit in the ones place. Let the digit in the ones place be x. This can be described by the equation: _____.

C. Questions that require solutions

7 In a football league, every team plays exactly one home game and one away game in a season. There are 182 games to play altogether in the season. How many teams are there in the league?

8 In 2017, a computer company generated a revenue of £6 million from the computer accessories department, which was 40% of the company's total revenue. The company forecast that its total revenue in 2019 would reach £21.6 million. If the annual growth rate from 2017 to 2019 is the same, what would the annual revenue in 2018 be?

9 A flower nursery occupies a rectangular plot of land of length 32 m and width 20 m. The owner is planning to build three paths, all of the same width (as shown in the diagram), to divide the nursery into 6 equal-sized fields for planting different types of tulip. Given that the total area of the six fields is 504 m², what is the width of the path?

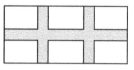

Diagram for question 9

10 A shed with area 130 m² is to be built. An existing wall is used as one side of the shed, as shown in the diagram. (the wall is 16 metres long.) A door of width 1 m is to be built on the side parallel to the wall. Given that the total length of the wooden panels used is 32 metres, find the length and width of the shed.

Diagram for question 10

Unit test 6

A. Multiple choice questions

1. Of these equations, (　　) is a quadratic equation in one variable x.

    A. $(x-1)(x+1) = (x-1)^2$　　　　B. $x^2 + \dfrac{1}{x^2} = 3$

    C. $(k+1)x^2 - 2kx + 1 = 0$　　　　D. $(m^2+2)x^2 - x + 2 = 0$

2. The quadratic equation with two roots $1+\sqrt{7}$ and $1-\sqrt{7}$ is (　　).

    A. $x^2 - 2x - 6 = 0$

    B. $x^2 - 2x + 6 = 0$

    C. $y^2 + 2y - 6 = 0$

    D. $y^2 + 2y + 6 = 0$

3. The solution to the equation $(x+1)(x-3) = 5$ is (　　).

    A. $x_1 = 1$ and $x_2 = -3$

    B. $x_1 = 4$ and $x_2 = -2$

    C. $x_1 = -1$ and $x_2 = 3$

    D. $x_1 = -4$ and $x_2 = 2$

4. The equation $x^2 + 2x - 3 = 0$ has (　　).

    A. no real roots

    B. two distinct real roots

    C. two equal real roots

    D. not sure

5. The revenue of a supermarket was £200 million in January and the total revenue in the first quarter was £1000 million. Given the monthly growth rate of the revenue was x, then we can establish an equation (　　).

    A. $200(1+x)^2 = 1000$

    B. $200 + 200 \times 2x = 1000$

    C. $200 + 200 \times 3x = 1000$

    D. $200[1 + (1+x) + (1+x)^2] = 1000$

B. Fill in the blanks

6 When m is _____ , the equation $(m - 3)x^{m-7} - x = 5$ in x is a quadratic equation in one variable. When m is _____ , the equation is a linear equation with one variable.

7 If $x^2 + x - 1 = 0$, then the value of the algebraic expression $x^3 + 2x^2 - 7$ is _____.

8 Given that the lengths of all the three sides of an isosceles triangle satisfy the equation $x^2 - 12x + 32 = 0$, then the perimeter of the triangle is _____.

9 Given that n $(n \neq 0)$ is a root of the equation $x^2 + mx + 2n = 0$ in x, then $m + n =$ _____.

10 Given that $2x^2 + 1$ and $4x^2 - 2x - 5$ sum to zero, then the value of x is _____.

11 Given that a is a real number and $\sqrt{(a - 4)} + \sqrt{(a^2 - 2a - 8)^2} = 0$, then the value of a is _____.

12 Given that the equations $3ax^2 - bx - 1 = 0$ and $ax^2 + 2bx - 5 = 0$ have a common root -1, then $a =$ _____ and $b =$ _____.

13 Given that $3 - \sqrt{2}$ is a root of the equation $x^2 + mx + 7 = 0$, then $m =$ _____ and the other root is _____.

14 Given that the two real roots of equation $x^2 - mx + 3 = 0$ are equal, then $m =$ _____.

15 The area of a rectangular field of grassland is 1000 m^2. If its length is 30 m longer than its width, then the width of the field of grassland is _____.

16 When $\sqrt{(a - 4)} + \sqrt{(b + 2)^2} + c^2 = 0$, then the solutions to the equation $ax^2 + bx + c = 0$ are _____.

17 Given that the equation in x, $x^2 + mx + n = 0$, has two equal real roots, then a set of real values of m and n could be $m =$ _____ and $n =$ _____.

C. Questions that require solutions

18 Factorise the expression $3x^2 + 12xy + 11y^2$.

19 Use a suitable method to solve each of these quadratic equations in one variable.

(a) $(3 - x)^2 + x^2 = 5$

(b) $4(x + 3)^2 = 25(x - 2)^2$

(c) $(2x + 3)^2 - 3(2x + 3) - 4 = 0$

(d) $x^2 - 2x + 1 - k(x - 1) = 0$

20 For what values of m, does the quadratic equation in one variable $(m^2 - 1)x^2 + 2(m - 1)x + 1 = 0$ satisfy these conditions?

(a) It has two distinct real roots.

(b) It has two equal real roots.

(c) It has no real roots.

21 Let the lengths of the three sides of $\triangle ABC$ be a, b and c. The equation $\frac{1}{2}x^2 + \sqrt{b}x + c - \frac{1}{2}a = 0$ in x has two equal real roots and the root of equation $3cx + 2b = 2a$ is $x = 0$.

(a) Determine the shape of $\triangle ABC$.

(b) Given that a and b are the two roots of the equation $x^2 + mx - 3m = 0$, find the value of m.

22 The two distinct real roots of the equation $a^2x^2 + (2a - 1)x + 1 = 0$ in x are x_1 and x_2.

(a) Find the set of values that a can take.

(b) Is there a real number a such that the two real roots of the equation sum to zero? If so, find the value of a; if not, explain why.

Look at Daniel's solution below.

(a) From the given, $\Delta = (2a-1)^2 - 4a^2 > 0$, and solving it for a, then $a < \frac{1}{4}$. So when $a < 0$, the equation has two distinct real roots.

(b) Yes, there is. If the two real roots of the equation sum to zero, then $x_1 + x_2 = -\frac{2a-1}{a^2} = 0$, so $a = \frac{1}{2}$. Therefore, when $a = \frac{1}{2}$, the two real roots of the equation, x_1 and x_2, add to zero.

Has Daniel made any mistakes? If so, identify them and correct them.

23 A factory plans to produce 1400 components in three months. It produces 200 components in the first month. Given that the monthly growth rate of the production for the next two months is the same, how much is the monthly growth rate?

24 The diagram shows a rectangle $ABCD$ with $AB = 6$ cm and $BC = 12$ cm. Point P moves from A towards B along AB with a speed of 1 cm/s. Point Q moves from B towards C along BC at a speed of 2 cm/s. Assume that P and Q start from A and B respectively at the same time.

(a) After how many seconds, is the area of $\triangle PBQ$ equal to 8 cm^2?

(b) After how many seconds, is the area of pentagon $APQCD$ the smallest? What is the smallest area of the pentagon?

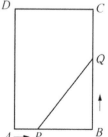

Diagram for question 24

Chapter 7   Quadratic equations in two variables

7.1   Introduction to quadratic equations in two variables

Learning objective

Understand the concepts of quadratic equations in two variables.

A. Multiple choice questions

1. Of these equations, the quadratic equation in two variables is (   ).

   A. $3x^2 = 1 - x$ B. $3x + y = 2$ C. $xy = y - x$ D. $\dfrac{1}{x^2} + y = 1$

2. Of these sets of simultaneous equations in two variables, (   ) consists of one linear and one quadratic equation.

   A. $\begin{cases} x = 1 \\ y^2 = 1 \end{cases}$ B. $\begin{cases} xy = 1 \\ \dfrac{1}{x} = 1 \end{cases}$ C. $\begin{cases} x^2 = y \\ \sqrt{x} = 1 \end{cases}$ D. $\begin{cases} x = 1 \\ 3x + y = 2 \end{cases}$

3. The set of solutions to the simultaneous equations $\begin{cases} x^2 = 1 \\ x + y = 2 \end{cases}$ is (   ).

   A. $\begin{cases} x = 1 \\ y = 1 \end{cases}$ and $\begin{cases} x = -1 \\ y = -1 \end{cases}$ B. $\begin{cases} x = 1 \\ y = 1 \end{cases}$ and $\begin{cases} x = -1 \\ y = 1 \end{cases}$

   C. $\begin{cases} x = 1 \\ y = 1 \end{cases}$ and $\begin{cases} x = -1 \\ y = -3 \end{cases}$ D. $\begin{cases} x = 1 \\ y = 1 \end{cases}$ and $\begin{cases} x = -1 \\ y = 3 \end{cases}$

B. Fill in the blanks

4. In a quadratic equation in two variables, there are _____ variables and the highest degree of the terms containing the variable(s) is _____.

5. In the equation $-5x^2 - 3xy + 4y^2 - \dfrac{x}{2} + y - 1 = 0$, the quadratic terms are _____, the coefficients of the linear terms are _____ and the constant is _____.

6. A quadratic equation in two variables that has a solution $\begin{cases} x = 1 \\ y = 4 \end{cases}$ could be _____.

C. Questions that require solutions

Convert the quadratic equations in two variables into two linear equations in two variables. (For example, by factorising $3xy - y^2$ into $y(3x - y)$, we can convert the quadratic equation $3xy - y^2 = 0$ into two linear equations $y = 0$, $3x - y = 0$.)

7 $x^2 + 2xy + y^2 = 9$

8 $x^2 - 3xy + 2y^2 = 0$

9 $(x - y)^2 - (x - y) - 2 = 0$

10 $(x - 2y)^2 - 3(x - 2y) - 28 = 0$

11 $2x^2 - xy - 3y^2 = 0$

12 Write three sets of solutions for the equation $x^2 = 2y + 1$.

Solve these sets of simultaneous equations.

13 $\begin{cases} x^2 = 4 \\ y^2 = 9 \end{cases}$

14 $\begin{cases} y^2 = 1 \\ x - y = 1 \end{cases}$

7.2 Solving simultaneous quadratic and linear equations

 Learning objective

Solve simultaneous equations involving linear and quadratic equations in two variables.

Solve these sets of simultaneous equations consisting of one linear and one quadratic equation in two variables.

1
$$\begin{cases} x^2 + y^2 = 13 \\ x + y = 5 \end{cases}$$

2
$$\begin{cases} y = x + 1 \\ x^2 - 2x + y = 3 \end{cases}$$

3
$$\begin{cases} 2x^2 - y^2 + x + 13 = 0 \\ 2x - y = 1 \end{cases}$$

4
$$\begin{cases} xy = -14 \\ x + y = 5 \end{cases}$$

5
$$\begin{cases} x - y = 1 \\ (x + 1)^2 + y^2 = 10 \end{cases}$$

6
$$\begin{cases} x + 5y = 4 \\ xy = -1 \end{cases}$$

7
$$\begin{cases} 2x - y = 1 \\ 10x^2 - y^2 - x + 1 = 0 \end{cases}$$

8
$$\begin{cases} x - y - 1 = 0 \\ 2x^2 - y^2 + x - 2y - 7 = 0 \end{cases}$$

9 $\begin{cases} x + y = 7 \\ x^2 + y^2 = 25 \end{cases}$

10 Given that the simultaneous equations $\begin{cases} x^2 + y^2 = 20 \\ x + y = m \end{cases}$ have exactly one set of real solutions, find the value of m and the solution to the simultaneous equations.

> A set of solutions to a pair of simultaneous equations gives the values of two variables that satisfy both equations.

11 The simultaneous equations $\begin{cases} y^2 = 4x \\ y = 2x + n \end{cases}$ have two sets of real different solutions.

(a) Find the set of values that n can take.

(b) Given that n is the greatest integer in the set, what are the solutions to the simultaneous equations?

7.3 Using quadratic (simultaneous) equations to solve application problems (1)

Learning objective

Use sets of simultaneous equations that include quadratic equations to solve these problems.

1 The revenue of a department store in February this year was 4 million pounds. The revenue in March increased by 10% compared to February. In May the revenue was 6.336 million pounds. Find the monthly growth rate of the revenue from March to May.

2 A shop purchased a batch of sunglasses at £21 each. If sunglasses are sold at £a per pair, the shop can sell $(350 - 10a)$ of them. However, due to a government regulation, the gross profit margin the shop can make from selling these sunglasses cannot exceed 20%. If the shop wants to make a profit of £400, then how many pairs of sunglasses should the shop purchase? What should the selling price of the sunglasses be?

3 Sami deposited £1000 in a one-year fixed deposit account. After one year, he withdrew the principle and the interest. He spent £500 of it, then he put the remaining amount into another one-year fixed deposit; at that time, the annual interest rate dropped by 10% compared to the previous interest rate. On maturity, the total sum of the principle and the interest was £530. Find the annual interest rate on the savings for his first deposit.

4 In a right-angled triangle, the length of the hypotenuse is 13 cm and the sum of the lengths of the other two sides is 17 cm. What is the area of the right-angled triangle?

5 The selling price of a jacket is £20. When the cost price was reduced by 50%, the shop's profit rate increased by 50%. What was the original cost price of the jacket? What was the shop's original profit rate?

7.4 Using quadratic (simultaneous) equations to solve application problems (2)

Learning objective

Use simultaneous equations involving quadratic equations to solve problems.

1. In a chess tournament, each contestant plays one game against every other contestant. The winner of each game scores 2 points, while the loser scores 0. If it is a tie, both contestants score 1 point. Four adjudicators have recorded the total scores of all the contestants and obtained four different results: 1979, 1980, 1984, and 1985. After verification, it is known that only one adjudicator's record is correct. Find the total number of contestants in the tournament.

2. A piece of wasteland, in the shape of a rectangle 18 metres long and 15 metres wide, is to be developed into garden, with paved regions taking up one-third (to the nearest 0.1 m^2) of the original area. Two designs have been suggested. The shading indicates the paved regions, in each case.

   (a) Design 1 (Diagram 1): The garden is divided into four sections by two paths, both the same width and perpendicular to each other.

   (b) Design 2 (Diagram 2): Four regions, all of the same size, are to be built: one at each corner.

   Can both the designs meet the requirement? If so, work out the width of the paths in Diagram 1 and the radius of the sectors in Diagram 2. If not, please give yours reasons.

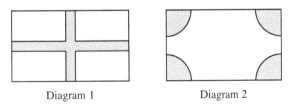

Diagram 1   Diagram 2

Diagram for question 2

③ The length of the hypotenuse of a right-angled triangle is 1 m longer than one of its shorter sides and the difference in length between the two perpendicular sides is 1 m. Find the length of the hypotenuse of the triangle.

④ A factory planned to produce 7200 tents for a disaster-hit area. Due to the high demand, the factory needed to produce 20% more tents than planned, and was asked to deliver all the tents to the area 5 days earlier than originally scheduled. Accordingly, the factory assigned more workers for the task so that the number of tents produced each day was twice that in the original plan. Given that the factory completed the task on schedule, how many tents did the factory actually produce each day?

⑤ Sharon bought a metres of pure cotton fabric and b metres of polyester fabric for £60. One year later, the prices changed. The price of pure cotton fabric was increased by a certain percentage, while the price of polyester fabric was reduced by exactly the same percentage. At the new prices, it would cost Sharon £36 to buy a metres of pure cotton fabric and £18 to buy b metres of polyester fabric. How much did it cost her to buy a metres of pure cotton fabric at the original price? By what percentage was the price of pure cotton fabric increased?

Unit test 7

A. Multiple choice questions

1 Look at these equations. The quadratic equation in two variables is (   ).

    A. $x^2 + x = 1$ B. $x^2 - 2y^2 = 5$ C. $x + 3yz = 0$ D. $\dfrac{1}{x^2} + \dfrac{1}{y^2} = 1$

2 $\begin{cases} x = 2 \\ y = -1 \end{cases}$ is a set of solutions to the simultaneous equations (   ).

    A. $\begin{cases} x^2 + 2x = 6 \\ x + 3y = -1 \end{cases}$ B. $\begin{cases} x^2 + 2x = 8 \\ x + 3y = -1 \end{cases}$ C. $\begin{cases} x^2 + 2x = 8 \\ x + 3y = 1 \end{cases}$ D. $\begin{cases} x^2 + 2x = 8 \\ x + 3y = 5 \end{cases}$

3 The solutions to the simultaneous equations $\begin{cases} x^2 = 4 \\ x - y = 0 \end{cases}$ are (   ).

    A. $\begin{cases} x = 4 \\ y = 4 \end{cases}$ and $\begin{cases} x = -4 \\ y = -4 \end{cases}$ B. $\begin{cases} x = 4 \\ y = -4 \end{cases}$ and $\begin{cases} x = -4 \\ y = 4 \end{cases}$

    C. $\begin{cases} x = 2 \\ y = 2 \end{cases}$ and $\begin{cases} x = -2 \\ y = -2 \end{cases}$ D. $\begin{cases} x = 2 \\ y = -2 \end{cases}$ and $\begin{cases} x = -2 \\ y = 2 \end{cases}$

4 For there to be one set of real number solutions to the system of equations $\begin{cases} 2x^2 + y^2 = 6 \\ mx + y = 3 \end{cases}$, the value of m must be (   ).

    A. $\pm\sqrt{3}$ B. $\pm\sqrt{2}$ C. ± 3 D. ± 1

B. Fill in the blanks

5 In the equation $x^2 - xy + \dfrac{x}{2} - 3y - 10 = 0$, the quadratic terms are _____ , the coefficients of the linear terms are _____ and the constant is _____ .

6 A quadratic equation in two variables that has a solution $\begin{cases} x = -1 \\ y = -2 \end{cases}$ could be _____ .

7 The solutions to the simultaneous equations $\begin{cases} x + y = 3 \\ (x-1)(y-3) = 0 \end{cases}$ are _____.

8 Given that, on a coordinate plane, point P is on the y-axis and its distance from point $(-3, 2)$ is 5, then the coordinates of point P are _____.

9 The quadratic equation $x^2 - 2xy - 3y^2 = 0$ can be converted into two linear equations, which are _____ and _____.

C. Questions that require solutions

10 Solve the simultaneous equations: $\begin{cases} x + y = 2 \\ x^2 - y^2 = 8 \end{cases}$.

11 Solve the simultaneous equations: $\begin{cases} x - 2y = 0 \\ 2x^2 + xy - 3y - 4 = 0 \end{cases}$.

12 Solve the simultaneous equations: $\begin{cases} x^2 - 2y^2 - 2x + y + 21 = 0 \\ 2x + y - 4 = 0 \end{cases}$.

13 Joshua uses 24 metres of wire to form a right-angled triangle. If he wants to make the hypotenuse 10 metres long, what will the lengths of the other two sides of the triangle be?

14 On one day, Alvin and Wade each deposited £10 000 into a two-year fixed savings account at two different banks. The annual interest rate on Wade's account is 25% higher than that on Alvin's. At the end of two years, the total interest Wade earned from his account was £209 more than the interest Avlin earned from his account. What were the annual interest rates of the two fixed savings accounts?

Chapter 8   Vectors

8.1   Basic concepts of vectors[*]

Learning objective

Understand the concepts of vectors, including terms and notation.

A. Multiple choice questions

1 Of these statements, the incorrect one is (   ).

A. A quantity that has both magnitude and direction is called a vector.

B. Two vectors are the same only when they have the same starting point.

C. Directed line segments of equal length, with different starting points but acting in the same direction, represent the same vector.

D. $|\overrightarrow{AB}| = |\overrightarrow{BA}|$

2 Of these statements, the correct one is (   ).

A. Only vectors acting in the same direction are parallel vectors.

B. If vectors $\overrightarrow{AB}$ and $\overrightarrow{BA}$ are parallel vectors, then they can be represented using symbols as $\overrightarrow{AB} = \overrightarrow{BA}$.

C. When two vectors are equal, the endpoints of the two directed line segments representing the two vectors are not necessarily the same.

D. When two vectors act in opposite directions, the endpoints of the two directed line segments are definitely not the same.

3 Of these expressions, (   ) is/are correct.

① If $|\mathbf{a}| = 0$, then $\mathbf{a} = 0$

② If $\mathbf{a} = \mathbf{0}$, then $|\mathbf{a}| = \mathbf{0}$;

③ If $|\mathbf{a}| = |\mathbf{b}|$, then $\mathbf{a} = \mathbf{b}$ or $\mathbf{a} = -\mathbf{b}$

④ If $\mathbf{a} = \mathbf{0}$, then $-\mathbf{a} = \mathbf{0}$

A. none

B. one

C. two

D. three

---

* In this chapter, we only deal with plane vectors, which are also called 2D vectors or simply vectors.

4　Let **b** be the inverse of **a**. The incorrect statement of the following is (　　).

　　A. The lengths of **a** and **b** must be equal.

　　B. **a** // **b**

　　C. The lines in which both **a** and **b** act must be parallel.

　　D. **a** is the inverse of **b**.

5　Of these statements, the incorrect one is (　　).

　　A. If **a** = **b** and **b** = **c**, then **a** = **c**.

　　B. If quadrilateral $ABCD$ is a parallelogram, then $\overrightarrow{AB} = \overrightarrow{CD}$.

　　C. If $\overrightarrow{AB} = \overrightarrow{CD}$, then $\overrightarrow{AB} = -\overrightarrow{DC}$.

　　D. If $\overrightarrow{AB} = \overrightarrow{CD}$, then $|\overrightarrow{AB}| = |\overrightarrow{CD}|$.

B. Fill in the blanks

6　If a vector is placed on a number line, with its starting point at the origin and the endpoint at 2, then its magnitude is _____ and the endpoint of the inverse vector also starting from the origin is _____.

7　In rectangle $ABCD$, diagonals AC and BD intersect at point O. Then the vector equal to $\overrightarrow{AO}$ is _____.

8　In the diagram, both quadrilaterals $ABCD$ and $ABDE$ are parallelograms.

　(a) The vectors equal to vector $\overrightarrow{AB}$ are _____.

　(b) If $|\overrightarrow{AB}| = 3$, then the magnitude of vector $\overrightarrow{EC}$ is _____.

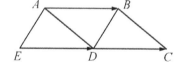

Diagram for question 8

C. Questions that require solutions

9　The diagram shows a parallelogram $ABCD$. All the segments in the diagram are drawn as directed line segments, and all the directed line segments are vectors.

　(a) Which pairs of vectors are equal?

　(b) In which pairs of vectors is one the inverse of the other?

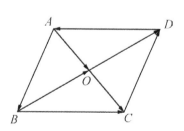

Diagram for question 9

10 In $\triangle ABC$, points D, E and F are the midpoints of sides AB, BC and CA, respectively.

    (a) Write down the vectors that are equal to $\overrightarrow{AD}$.

    (b) Write down the inverse of $\overrightarrow{DE}$.

    (c) How many non-zero vectors are parallel to $\overrightarrow{FE}$?

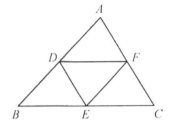

Diagram for question 10

11 In the diagram, point O is the centre of regular hexgon $ABCDEF$. Identify:

    (a) the vectors that are equal to $\overrightarrow{BA}$

    (b) the vectors that are the inverse of $\overrightarrow{AO}$

    (c) the number of non-zero vectors that are parallel to $\overrightarrow{DE}$.

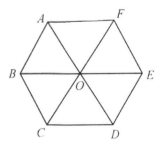

Diagram for question 11

8.2 Addition of vectors (1)

Learning objective

Know how to add vectors.

A. Multiple choice questions

1 Of these equations, the incorrect one is (　　).

A. $\mathbf{0} + \mathbf{0} = \mathbf{0}$

B. $\mathbf{a} + -\mathbf{a} = \mathbf{0}$

C. $\mathbf{a} + \mathbf{0} = \mathbf{0} + \mathbf{a} = \mathbf{a}$

D. $((\mathbf{a} + \mathbf{b}) + \mathbf{c}) + \mathbf{c} = (\mathbf{a} + \mathbf{c}) + \mathbf{b}$

2 Of these statements about zero vectors, the correct one is (　　).

A. A zero vector has no length.

B. A zero vector has no direction.

C. A zero vector is parallel to any vector.

D. A zero vector is not parallel to any vector.

3 Look at the diagram. If in parallelogram $ABCD$, $\overrightarrow{AB} = \mathbf{a}$ and $\overrightarrow{AD} = \mathbf{b}$, then (　　) is correct.

A. $\mathbf{a} + \mathbf{b} = \overrightarrow{DB}$

B. $\mathbf{a} + \mathbf{b} = \overrightarrow{CA}$

C. $\mathbf{a} + \mathbf{b} = \overrightarrow{AC}$

D. $\mathbf{a} + \mathbf{b} = \overrightarrow{BD}$

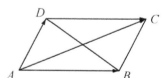

Diagram for question 3

B. Fill in the blanks

4 When adding vectors, the general rule we follow is called the _____, that is, $\overrightarrow{AB} + \overrightarrow{BC} =$ _____ and the sum of two or more vectors is called a _____ vector.

5 In general, we call a vector of zero length a _____ and denote it as _____.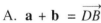

6 If **a** represents 1 km southwards, **b** represents 2 km southwards and **c** represents 3 km northwards, then **a** + **b** represents _____, **a** + **c** represents _____, and **a** + **b** + **c** represents _____.

7 The diagram shows a parallelogram $ABCD$, with diagonals AC and BD intersecting at O. Complete the statements of additions of vectors.

(a) $\overrightarrow{AB} + \overrightarrow{BC} = $ _____;

(b) $\overrightarrow{AC} + \overrightarrow{CD} + \overrightarrow{DO} = $ _____;

(c) $\overrightarrow{AC} + \overrightarrow{CD} + \overrightarrow{DA} = $ _____;

(d) $\overrightarrow{AB} + \overrightarrow{BC} + \overrightarrow{CD} = $ _____.

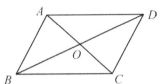

Diagram for question 7

8 The diagram shows a parallelogram $ABCD$, with E being the midpoint of BC.

(a) **Draw** the vector sum of $\overrightarrow{AD}$ and $\overrightarrow{DC}$: $\overrightarrow{AD} + \overrightarrow{DC} = $ _____.

(b) **Draw** the vector sum of $\overrightarrow{AD}$ and $\overrightarrow{EA}$: $\overrightarrow{AD} + \overrightarrow{EA} = $ _____.

(c) If all the line segments in the diagram **are directed line segments**, then of all the vectors represented by those directed line segments, the vectors that are the inverse of $\overrightarrow{BE}$ are _____.

(d) $\overrightarrow{AB} + \overrightarrow{BE} + \overrightarrow{EA} = $ _____.

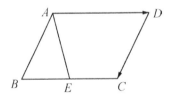

Diagram for question 8

 C. Questions that require solutions

9 The diagram shows the vectors **a**, **b** and **c** and **d** (|**a**| + |**c**| = |**d**|). Construct these vectors. (Simply draw them, there is no need to write the steps.)

(a) **a** + **b**   (b) **a** + **c**   (c) **a** + **d**

(d) **a** + **b** + **c**   (e) **a** + **c** + **d**

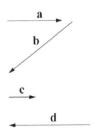

Diagram for question 9

10 Two vectors $\overrightarrow{AB}$ and $\overrightarrow{CD}$ are parallel. Place them on a number line so that point A is at the origin, point B is at 2 and point C is at 10.

Given that $\overrightarrow{AB} + \overrightarrow{CD} = \overrightarrow{EF}$, E is at the origin and F is at 1, where is point D?

11 The diagram shows a right-angled triangle ABC, with $\angle ACB$ being the right angle. CD is the perpendicular height of C from hypotenuse AB, $|\overrightarrow{AC}| = 4$, and $|\overrightarrow{BC}| = 3$.

Find: (a) $|\overrightarrow{CD}|$ (b) $|\overrightarrow{AC} + \overrightarrow{CB}|$

(c) $|\overrightarrow{AC} + \overrightarrow{BC}|$ (d) $|\overrightarrow{BC} + \overrightarrow{CD} + \overrightarrow{DA}|$.

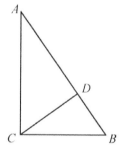

Diagram for question 11

8.3 Addition of vectors (2)

Learning objective

Use diagrams to add vectors and represent vector addition.

A. Multiple choice questions

1 Given that $\overrightarrow{CD}$ is a non-zero vector, then the correct equation is (   ).

   A. $\overrightarrow{CD} + \overrightarrow{DC} = 0$ B. $|\overrightarrow{CD}| + |\overrightarrow{DC}| = 0$

   C. $|\overrightarrow{CD}| = |\overrightarrow{DC}|$ D. $|\overrightarrow{CD}| = -|\overrightarrow{DC}|$

2 In quadrilateral $ABCD$, if $\overrightarrow{AC} = \overrightarrow{AB} + \overrightarrow{AD}$ and $|\overrightarrow{AB}| = |\overrightarrow{AD}|$, then quadrilateral $ABCD$ is
(   ).

   A. a rectangle   B. a rhombus

   C. a square   D. neither a rectangle nor a rhombus

B. Fill in the blanks

3 In the diagram:

  (a) $\overrightarrow{AC} + \overrightarrow{CB} =$ _____   (b) $\overrightarrow{AC} + \overrightarrow{CD} + \overrightarrow{DE} =$ _____ .

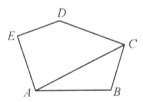

Diagram for question 3

4 In the diagram, quadrilateral $ABCD$ is a square with side length 1. Points M and N are the midpoints of AD and BC, respectively. Then $|\overrightarrow{BM} + \overrightarrow{CD} + \overrightarrow{MN} + \overrightarrow{MD}| =$ _____ .

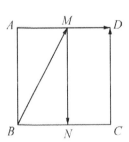

Diagram for question 4

5 In the diagram, point O is the centre of the regular hexagon.

Then $\overrightarrow{AB} + \overrightarrow{CD} + \overrightarrow{EF} + \overrightarrow{OB} + \overrightarrow{OD} + \overrightarrow{OF} = $ _____ .

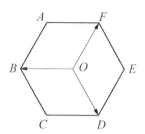

Diagram for question 5

6 The diagram shows $\triangle ABC$; points D, E and F are the midpoints of the three sides. On the diagram, draw and work out each sum of vectors.

(a) $\overrightarrow{AB} + \overrightarrow{BC} = $ _____

(b) $\overrightarrow{DE} + \overrightarrow{BD} = $ _____

(c) $\overrightarrow{FB} + \overrightarrow{DC} = $ _____

(d) $\overrightarrow{BA} + \overrightarrow{FD} = $ _____

(e) $\overrightarrow{BD} + \overrightarrow{CE} + \overrightarrow{FE} = $ _____ .

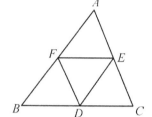

Diagram for question 6

C. Questions that require solutions

7 The diagram shows vectors **a**, **b**, **c**, **d** and **e**. Construct $\mathbf{a} + \mathbf{b} + \mathbf{c} + \mathbf{d} + \mathbf{e}$.

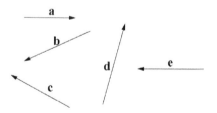

Diagram for question 7

8 Given that **a** represents a move of 3 km eastwards, **b** represents a move of 2 km southwards, **c** represents a move of 1 km westwards and **d** represents a move of 4 km northwards, then what does $\mathbf{a} + \mathbf{b} + \mathbf{c} + \mathbf{d}$ represent?

9 The diagram shows a rhombus $ABCD$, point O is the intersection point of the diagonals and $|\overrightarrow{BA} + \overrightarrow{BC} + \overrightarrow{DO}| : |\overrightarrow{CB} + \overrightarrow{CD} + \overrightarrow{AO}| = \sqrt{3}$. Find the sizes of the interior angles of rhombus $ABCD$.

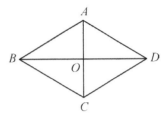

Diagram for question 9

8.4 Subtraction of vectors (1)

Learning objective

Know how to subtract vectors.

A. Multiple choice questions

1. Look at the diagram. The correct equation is (   ).

   A. $\overrightarrow{DA} - \overrightarrow{DC} = \overrightarrow{AC}$

   B. $\overrightarrow{DA} - \overrightarrow{CD} = \overrightarrow{CA}$

   C. $\overrightarrow{BA} - \overrightarrow{BC} = \overrightarrow{AC}$

   D. $\overrightarrow{BC} - \overrightarrow{BA} = \overrightarrow{AC}$

   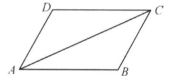

   Diagram for question 1

2. The diagram shows the vectors **a** and **b**. Of these diagrams, the one representing the vector **a** – **b** correctly is (   ).

   Diagram for question 2

   A. $\overrightarrow{BA}$ (b a B O A)

   B. $\overrightarrow{BA}$ (b a O B A)

   C. $\overrightarrow{BO}$ (a b O A B)

   D. $\overrightarrow{OB}$ (a b O A B)

3. Look at these statements about **a** + **b** and **a** – **b**. The correct one is (   ).

   A. $\mathbf{a} + \mathbf{b} > \mathbf{a} - \mathbf{b}$

   B. $|\mathbf{a} + \mathbf{b}| > |\mathbf{a} - \mathbf{b}|$

   C. **a** + **b** and **a** – **b** are parallel vectors.

   D. **a** + **b** could be the inverse of **a** – **b**.

4. In quadrilateral $ABCD$, if $\overrightarrow{AB}$ = **a**, $\overrightarrow{AD}$ = **b** and $\overrightarrow{BC}$ = **c**, then $\overrightarrow{CD}$ equals (   ).

   A. $\mathbf{a} - \mathbf{b} + \mathbf{c}$ B. $-\mathbf{a} + \mathbf{b} - \mathbf{c}$

   C. $\mathbf{a} - \mathbf{b} - \mathbf{c}$ D. $-\mathbf{a} + \mathbf{b} + \mathbf{c}$

B. Fill in the blanks

5 Calculate: $\overrightarrow{OA} - \overrightarrow{OB} = $ _____ , $\overrightarrow{AD} - \overrightarrow{BD} = $ _____ .

6 The diagram shows quadrilateral $ABCD$.

$\overrightarrow{BA} + \overrightarrow{AD} - \overrightarrow{CD} = $ _____ $\overrightarrow{BD} - \overrightarrow{BC} = $ _____

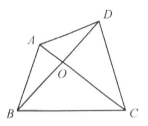

Diagram for question 6

7 Express these vectors in terms of **a**, **b** and **c**, as shown in the figure.

(a) $\overrightarrow{CD} = $ _____

(b) $\overrightarrow{BD} = $ _____

(c) $\overrightarrow{CA} = $ _____

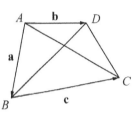

Diagram for question 7

8 The diagram shows a right-angled isosceles triangle ABC, with $\angle C = 90°$ and $AC = BC = 2$ cm. Points M and N are the midpoints of AC and BC, respectively.

Then $| \overrightarrow{BA} - \overrightarrow{BM} - \overrightarrow{NA} | = $ _____ .

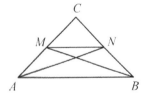

Diagram for question 8

C. Questions that require solutions

9 The diagram shows vectors **a**, **b** and **c**. Construct these vectors.

(a) **a** + **b**   (b) **b** – **a**   (c) **a** – (**b** – **c**)

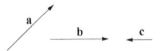

Diagram for question 9

10 Are these equations correct? If not, make the necessary correction.

(a) $\overrightarrow{CB} - \overrightarrow{AB} = \overrightarrow{CA}$

(b) $\overrightarrow{CB} - \overrightarrow{AB} - \overrightarrow{CA} = 0$

11 Simplify each equation.

(a) $\overrightarrow{AB} - \overrightarrow{AC} + \overrightarrow{BD} - \overrightarrow{CD}$

(b) $-\overrightarrow{OB} + \overrightarrow{AD}$

(c) $\overrightarrow{NQ} + \overrightarrow{QP} + \overrightarrow{MN} - \overrightarrow{MP}$

12 Given that quadrilateral $ABCD$ is a rectangle, $AB = 1$ and $BC = 2$, find:

(a) $|\overrightarrow{AB} - \overrightarrow{CB}|$

(b) $|\overrightarrow{AB} - \overrightarrow{BC} + \overrightarrow{AC}|$

(c) $\overrightarrow{AB} - \overrightarrow{CB} - \overrightarrow{AC}$

(d) $\overrightarrow{AC} + \overrightarrow{BD} - \overrightarrow{BC}.$

8.5 Subtraction of vectors (2)

Learning objective

Use diagrams to subtract vectors and represent vector subtraction.

A. Multiple choice questions

1. Given a parallelograms $ABCD$, if the diagonals AC and BD intersect at point O, $\overrightarrow{AB} = \mathbf{a}$ and $\overrightarrow{BC} = \mathbf{b}$, then $\overrightarrow{CO}$ can be represented as (   ).

A. $\frac{1}{2}(\mathbf{a} + \mathbf{b})$ B. $-\frac{1}{2}(\mathbf{a} + \mathbf{b})$ C. $\frac{1}{2}(\mathbf{a} - \mathbf{b})$ D. $\frac{1}{2}(\mathbf{b} - \mathbf{a})$

2. The diagram shows a parallelogram $ABCD$. Points E, F, G and H are the midpoints of the four sides. There are (   ) vectors that are equal to vector $\overrightarrow{OF}$.

A. 3   B. 5

C. 7   D. 9

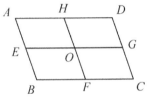

Diagram for question 2

3. The diagram shows a parallelogram $ABCD$. The correct equation is (   ).

A. $\overrightarrow{AB} + \overrightarrow{CD} = \overrightarrow{AD} + \overrightarrow{BD}$

B. $\overrightarrow{AB} + \overrightarrow{CD} = \overrightarrow{AC} - \overrightarrow{BD}$

C. $\overrightarrow{AB} - \overrightarrow{CD} = \overrightarrow{AC} + \overrightarrow{BD}$

D. $\overrightarrow{AB} - \overrightarrow{CD} = \overrightarrow{AC} - \overrightarrow{BD}$

Diagram for question 3

B. Fill in the blanks

4. In the diagram, quadrilateral $ABCD$ is a parallelogram. Expressed in terms of $\mathbf{a}$, $\mathbf{b}$ and $\mathbf{c}$,

$\overrightarrow{AC} = $ _____ = _____

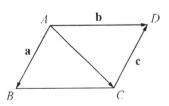

Diagram for question 4

5 In the diagram, $\triangle ABC$ is an isosceles triangle, $AB = AC = 3$ cm and $BC = 2$ cm. Points D, E and F are the midpoints of BC, CA and AB, respectively.

Then: $|\overrightarrow{DE} - \overrightarrow{DF}| = $ _____

$|\overrightarrow{DE} + \overrightarrow{DF}| = $ _____ .

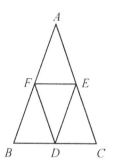

Diagram for question 5

6 In the diagram, quadrilaterals $ABOF$, $BCDO$ and $DEFO$ are all parallelograms. Expressing the following vectors in terms of $\mathbf{a}$, $\mathbf{b}$ and $\mathbf{c}$,

$\overrightarrow{AF} = $ _____ $\overrightarrow{OC} = $ _____ $\overrightarrow{DF} = $ _____

$\overrightarrow{AE} = $ _____ $\overrightarrow{CF} = $ _____ .

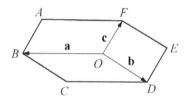

Diagram for question 6

7 Given that the side length of square $ABCD$ is 1, $\overrightarrow{AB} = \mathbf{a}$, $\overrightarrow{AC} = \mathbf{c}$ and $\overrightarrow{BC} = \mathbf{b}$, then $|\mathbf{a} + \mathbf{b} + \mathbf{c}| = $ _____ .

C. Questions that require solutions

8 Given the vectors $\mathbf{a}$, $\mathbf{b}$, $\mathbf{c}$ and $\mathbf{d}$ as shown in the diagram, draw these vectors.

(a) $\mathbf{a} + \mathbf{c} - \mathbf{b} - \mathbf{d}$ (b) $(\mathbf{a} - \mathbf{c}) + (\mathbf{b} - \mathbf{d})$

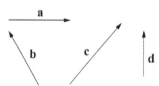

Diagram for question 8

9 The diagram shows a parallelogram $OBCA$. Point D is on OB.

(a) Write your answers in the blanks.

$\overrightarrow{OA} + \overrightarrow{AC} = $ _____ $\overrightarrow{AD} - \overrightarrow{OD} = $ _____

The vectors parallel to $\overrightarrow{AC}$ are _____ .

(b) Draw: $\overrightarrow{OA} + \overrightarrow{CD} - \overrightarrow{AD}$.

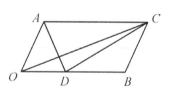

Diagram for question 9

10 The diagram shows a trapezium $ABCD$, in which $AD \parallel BC$, and point E is on BC. Connect DE and AC.

(a) Write your answers in the blanks.

$\overrightarrow{CD} + \overrightarrow{DE} =$ _____ $\overrightarrow{BC} - \overrightarrow{BA} =$ _____

(b) If all the line segments in the diagram are **drawn as directed line segments**, write down four vectors that are parallel to $\overrightarrow{BE}$.

(c) Draw: $\overrightarrow{AB} + \overrightarrow{AD}$. (Indicate which is the resultant vector.)

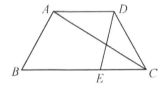

Diagram for question 10

11 In $\triangle ABC$, point D is the midpoint of AB and point E is a point on the extension of BC beyond C. $BE = 2BC$.

(a) Use vector $\overrightarrow{BC}$ to express vector $\overrightarrow{BE}$.

(b) Use $\overrightarrow{BA}$ and $\overrightarrow{BC}$ to express vector $\overrightarrow{DE}$.

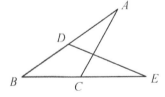

Diagram for question 11

8.6 Multiplying a vector by a scalar (1)

Learning objective

Know how to multiply vectors by scalar numbers.

A. Multiple choice questions

1 Of these linear operations on vectors, the incorrect one is (   ).

A. $\mathbf{a} + \mathbf{d} = \mathbf{d} + \mathbf{a}$

B. $(\mathbf{a} + \mathbf{b}) + \mathbf{c} = \mathbf{a} + (\mathbf{b} + \mathbf{c})$

C. $\mathbf{a}(\mathbf{b} + \mathbf{c}) = \mathbf{ab} + \mathbf{ac}$

D. $(-\mathbf{b}) + (-\mathbf{b}) + (-\mathbf{b}) + (-\mathbf{b}) = -4\mathbf{b}$

> Linear operations on vectors include addition, subtraction, and multiplication of vectors by a scalar (i.e., a real number).

2 The diagram shows a parallelogram $ABCD$. Points E, F, G and H are the midpoints of the four sides, EG intersects FH at point O. Given $\overrightarrow{AD} = \mathbf{a}$ and $\overrightarrow{CD} = \mathbf{b}$, then (   ) of these equations is/are correct.

① $\overrightarrow{FA} = \dfrac{1}{2}\mathbf{b}$ ② $\overrightarrow{OF} = -\dfrac{1}{2}\mathbf{a}$ ③ $\overrightarrow{BA} + \overrightarrow{HD} = \dfrac{3}{2}\mathbf{b}$

④ $\overrightarrow{BC} + \overrightarrow{EA} = \dfrac{3}{2}\mathbf{a}$

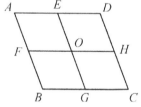

Diagram for question 2

A. one   B. two   C. three   D. four

3 Read these statements about $\mathbf{a}$ and $\mathbf{b}$.

① $\mathbf{a} /\!/ \mathbf{b}$ ② $\mathbf{a} = -\mathbf{b}$ ③ $\mathbf{a} + \mathbf{b} = 0$ ④ $|\mathbf{a}| = |\mathbf{b}|$

Given that $\mathbf{a}$ is the inverse of $\mathbf{b}$, then (   ) of these equations is/are correct.

A. one   B. two   C. three   D. four

B. Fill in the blanks

4 If $k = 0$ or $\mathbf{a} = \mathbf{0}$, then $k\mathbf{a} = $ _____ .

5 Given the non-zero vector $\overrightarrow{OA} = \mathbf{a}$, vector $\overrightarrow{OB}$ acts in the same direction as $\mathbf{a}$ and its length is 6 times $|\mathbf{a}|$, then $\overrightarrow{OB} = $ _____ .

6 Given the non-zero vector $\overrightarrow{OC} = \mathbf{c}$, vector $\overrightarrow{OD}$ acts in the opposite direction of $\mathbf{c}$ and its length is $\frac{1}{5}$ of $|\mathbf{c}|$, then $\overrightarrow{OD} = $ _____.

7 Three non-zero vectors $\overrightarrow{OM}$, $\overrightarrow{ON}$ and $\overrightarrow{OT}$ act in the same direction, $\overrightarrow{OM} = \mathbf{a}$, $\overrightarrow{ON} = \mathbf{b}$, and the length of $|\overrightarrow{OT}|$ is the sum of m times $|\mathbf{a}|$ and n times $|\mathbf{b}|$ (m, n are real numbers). Then $\overrightarrow{OT}$ expressed in terms of $\mathbf{a}$ and $\mathbf{b} = $ _____.

8 In the diagram, all the sides of $\triangle ABC$ are directed line segments. There are in total _____ directed vectors in the diagram and they are _____.

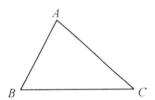

Diagram for question 8

9 Given two real numbers m and n, which are not both zero at the same time, if $(m \cdot n)\mathbf{a} = \mathbf{0}$, then _____.

C. Questions that require solutions

10 Given a non-zero vector $\mathbf{a}$, as shown in the diagram, construct $\frac{4}{3}\mathbf{a}$.

$$\overrightarrow{\qquad\mathbf{a}\qquad}$$

Diagram for question 10

11 The diagram shows a trapezium $ABCD$ with $AB \parallel DC$. Points F and E are on AD and BC, respectively, and $FE \parallel DC$. Given that $5AB = 3CD$ and $CE = 2BE$, use vector $\overrightarrow{DC}$ to express vector $\overrightarrow{FE}$.

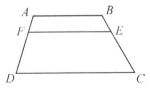

Diagram for question 11

12 Given non-zero vectors **a** and **b**, as shown in the diagram, construct $\sqrt{5}\,\mathbf{b} - \dfrac{2}{3}\mathbf{a}$.

Diagram for question 12

8.7 Multiplying a vector by a scalar (2)

Learning objective

Perform scalar multiplication of vectors and represent the solutions using diagrams.

A. Multiple choice questions

1. Of the following linear operations on vectors, the incorrect one is (   ).

   A. If c and d are real numbers, then $c(d\mathbf{a}) = (cd)\mathbf{a}$.

   B. If c is real number, then $c(\mathbf{a} + \mathbf{b}) = c\mathbf{a} + c\mathbf{b}$.

   C. If c and d are real numbers, then $(c + d)\mathbf{a} = c\mathbf{a} + d\mathbf{a}$.

   D. If c and d are real numbers and $cd\mathbf{a} = \mathbf{0}$, then $\mathbf{a} = 0$.

2. Of the following statements, the incorrect one is (   ).

   A. If m identical vectors $\mathbf{b}$ are added to gether, the result is $m\mathbf{b}$)

   B. When $k > 0$ and $\mathbf{a} = \mathbf{0}$, vectors $k\mathbf{a}$ and $\mathbf{a}$ act in the same direction.

   C. If $k\mathbf{a}$ and $\mathbf{a}$ are parallel, then the two vectors act in the same direction.

   D. When $k > 0$, $|k\mathbf{a}| = k|\mathbf{a}|$.

3. Of the following statements, (   ) are correct.

   ① Let k be a real number and $\mathbf{a}$ be a vector, then the product of k and $\mathbf{a}$ is a vector.

   ② If $k \neq 0$ and $\mathbf{a} \neq \mathbf{0}$, then the length of $k\mathbf{a}$ is $|k||\mathbf{a}|$.

   ③ If $k = 0$ or $\mathbf{a} = \mathbf{0}$, then $k\mathbf{a} = \mathbf{0}$.

   ④ If $k > 0$, then the direction of $k\mathbf{a}$ is opposite to the direction of $\mathbf{a}$.

   A. 1   B. 2

   C. 3   D. 4

4. Let m and n be real numbers. Then the incorrect equation is (   ).

   A. $m(n\mathbf{a}) = (mn)\mathbf{a}$

   B. $(m + n)\mathbf{a} = m\mathbf{a} + n\mathbf{a}$

   C. $m(\mathbf{a} + \mathbf{b}) = m\mathbf{a} + m\mathbf{b}$

   D. $m \times \mathbf{0} = 0$

B. Fill in the blanks

5 Calculate: $12(5\mathbf{a})$ = _____ , based on the _____ law of scalar multiplication of vectors.

6 Calculate: $5(\mathbf{a} + \mathbf{b})$ = _____ , based on the _____ law of scalar multiplication over addition of vectors.

7 Calculate: $2(\mathbf{a} - \mathbf{b}) + 3\mathbf{b}$ = _____ .

8 Calculate: $3m\mathbf{a} - n\mathbf{a} + (-2m\mathbf{a})$ = _____ .

9 Calculate: $\left(\mathbf{a} - \dfrac{3}{2}\mathbf{b}\right) - 5(3\mathbf{a} - 2\mathbf{b})$ = _____ .

10 Calculate: $7(\mathbf{a} - 2\mathbf{c} + 2\mathbf{b}) + 4(\mathbf{a} - 3\mathbf{b} + \mathbf{c}) - 8(\mathbf{a} - 2\mathbf{c})$ = _____ .

C. Questions that require solutions

11 Given that vectors $\mathbf{a}$, $\mathbf{b}$ and $\mathbf{x}$ satisfy the relation $2\mathbf{b} - 3(\mathbf{a} - 3\mathbf{b} - 2\mathbf{x}) = \mathbf{0}$, express $\mathbf{x}$ in terms of vectors $\mathbf{a}$ and $\mathbf{b}$.

12 The diagram shows vectors $\mathbf{a}$ *and* $\mathbf{b}$. *Construct* $\dfrac{1}{3}(2\mathbf{a} - 3\mathbf{b})$.

Diagram for question 12

13 Given non-zero real numbers m and n ($m \neq n$) and non-zero vectors $\mathbf{a}$, $\mathbf{b}$ and $\mathbf{x}$, which satisfy the relation $(m - n)(2\mathbf{a} - 3\mathbf{x}) = 4\mathbf{b}$, then use vectors $\mathbf{a}$ and $\mathbf{b}$ and real numbers m and n to represent vector $\mathbf{x}$.

8.8 Multiplying a vector by a scalar (3)

Learning objective

Understand the concept of a unit vector; use scalar multiplication of vectors to solve problems.

A. Multiple choice questions

1. Of these statements, the incorrect one is (   ).

   A. A vector of length 1 is called a unit vector.

   B. If $\mathbf{e}$ is a unit vector, then $|\mathbf{e}| = 1$.

   C. $\mathbf{a}$ and $\mathbf{b}$ are two unit vectors.

   D. If $\mathbf{a} = m\mathbf{b}$, then $\mathbf{a} /\!/ \mathbf{b}$.

2. Of these statements, the correct one is (   ).

   A. If non-zero vectors $\mathbf{a}$ and $\mathbf{b}$ are such that $\mathbf{a} /\!/ \mathbf{b}$, then $\mathbf{a}$ and $\mathbf{b}$ act in the same direction.

   B. If non-zero vectors $\mathbf{a}$ and $\mathbf{b}$ are such that $\mathbf{a} /\!/ \mathbf{b}$, then $\mathbf{a}$ and $\mathbf{b}$ act in opposite directions.

   C. If non-zero vectors $\mathbf{a}$ and $\mathbf{b}$ are such that $\mathbf{a} /\!/ \mathbf{b}$, then there exists one and only one real number m such that $\mathbf{b} = m\mathbf{a}$.

   D. If non-zero vectors $\mathbf{a}$ and $\mathbf{b}$ are such that $\mathbf{a} = m\mathbf{b}$, then the lines that both $\mathbf{a}$ and $\mathbf{b}$ are on are parallel.

3. Of these, the correct one is (   ).

   A. For non-zero vectors $\mathbf{a}$, $\mathbf{b}$, and $\mathbf{c}$, if $\mathbf{a} /\!/ \mathbf{c}$, $\mathbf{b} /\!/ \mathbf{c}$, then $\mathbf{a} /\!/ \mathbf{b}$.

   B. Given $\mathbf{e}$ is a unit vector, then $\mathbf{e} = 1$.

   C. For any non-zero vector $\mathbf{a}$, if a unit vector acting in the same direction as $\mathbf{a}$ is $\mathbf{a}_0$, then $\mathbf{a} = \mathbf{a}_0$.

   D. A zero vector has no magnitude but it has infinitely many directions.

4. C is the midpoint of segment AB and $\overrightarrow{AB} = \mathbf{a}$. Then the correct equation is (   ).

   A. $\overrightarrow{AC} = \dfrac{1}{2}\mathbf{a}$ B. $\overrightarrow{BC} = \dfrac{1}{2}\mathbf{a}$ C. $\overrightarrow{AC} = \overrightarrow{BC}$ D. $\overrightarrow{AC} + \overrightarrow{BC} = 0$

B. Fill in the blanks

5 Given non-zero vector **a** and non-zero real number m, such that $\mathbf{a} = m\mathbf{b}$, then **a** _____ **b**. (Choose "is parallel to" or "is not parallel to".)

6 Given two different vectors $\overrightarrow{AB}$ and $\overrightarrow{CD}$, such that $2\overrightarrow{AB} = 3\overrightarrow{CD}$, then the relation between segments AB and CD in terms of their positions is _____ and the relation between segments AB and CD in terms of length is _____.

7 Given that $\mathbf{b} = k\mathbf{a}$, $|\mathbf{a}| = 2$ and $|\mathbf{b}| = 6$, then the real number k = _____.

8 For a real number $k \neq 0$, when non-zero vectors **b** and **a** act in opposite directions and $k|\mathbf{b}| = |\mathbf{a}|$, then $\mathbf{b} =$ _____.

9 Given that **e** is a unit vector, **a** acts in the same direction as **e** and its length is 8, then $\mathbf{a} =$ _____ **e**.

10 Given vectors **a**, **b**, **c**, $\mathbf{a} + \mathbf{b} = 2\mathbf{c}$ such that $\mathbf{a} - \mathbf{b} = 3\mathbf{c}$, then vectors **a** and **b** are parallel and act _____ . (Choose "in the same direction" or "in opposite directions".)

C. Questions that require solutions

11 For vectors **a**, **b** and **c**, $\mathbf{a} + 2\mathbf{b} = 3\mathbf{c}$ and $\mathbf{b} + \dfrac{20}{3}\mathbf{c} = 2\mathbf{a}$. Use a suitable method to eliminate **a**, to determine whether vectors **b** and **c** are parallel but in opposite directions. Explain your conclusion, referring to the meanings of scalars and vectors.

12 The diagram shows a right-angled trapezium $ABCD$ with $AD \parallel BC$ and $\angle ABC = 90°$. E and F are the midpoints of AB and CD, respectively. $AD = 3$, $BC = 6$ and $AB = 4$.

Let $\overrightarrow{AD} = \mathbf{a}$. Express vectors $\overrightarrow{CB}$ and $\overrightarrow{EF}$ in terms of $\mathbf{a}$.

Let $\overrightarrow{AB} = \mathbf{b}$. Express $\overrightarrow{DC}$ in terms of $\mathbf{a}$ and $\mathbf{b}$.

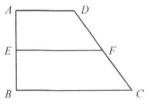

Diagram for question 12

13 The diagram shows a rectangular window made from eight identical glass rectangles. $AH = 30$ cm, $AT = 20$ cm, $\overrightarrow{AT} = \mathbf{a}$, and $\overrightarrow{AH} = \mathbf{b}$.

(a) Express vectors $\overrightarrow{AB}$, $\overrightarrow{DC}$ and $\overrightarrow{BC}$ in terms of $\mathbf{a}$ and $\mathbf{b}$.

(b) Simplify $\overrightarrow{AB} + \overrightarrow{BC} + \overrightarrow{DC} - \overrightarrow{DA}$ (in terms of $\mathbf{a}$ and $\mathbf{b}$).

(c) Express $\overrightarrow{HE}$ and $\overrightarrow{EF}$ in terms of $\mathbf{a}$ and $\mathbf{b}$.

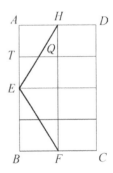

Diagram for question 13

8.9 Linear operations on vectors (1)

Learning objective

Apply linear operations on vectors; use vectors to construct geometric arguments and proofs.

A. Multiple choice questions

1 Of these linear operations, the incorrect one is (   ).

A. $2(\mathbf{a}+2\mathbf{b})-3(\mathbf{a}-\mathbf{b}) = -\mathbf{a}+7\mathbf{b}$

B. $\frac{1}{2}(\mathbf{a}+2\mathbf{b})+2\left(\mathbf{a}-\frac{1}{2}\mathbf{b}\right) = \frac{5}{2}\mathbf{a}$

C. $(\mathbf{a}+2\mathbf{b})-2(\mathbf{b}-\mathbf{a}) = 3\mathbf{a}$

D. $3(2\mathbf{b}-3\mathbf{a})-(\mathbf{a}-2\mathbf{b}) = -8\mathbf{a}+10\mathbf{b}$

2 In rectangle $ABCD$, O is the midpoint of AC. Given that $\overrightarrow{BC} = 3\mathbf{a}$ and $\overrightarrow{DC} = 2\mathbf{b}$, then $\overrightarrow{AO}$ is (   ).

A. $\frac{1}{2}(3\mathbf{a}+2\mathbf{b})$
B. $\frac{1}{2}(3\mathbf{a}-2\mathbf{b})$
C. $\frac{1}{2}(2\mathbf{b}-3\mathbf{a})$
D. $\frac{1}{2}(3\mathbf{b}+2\mathbf{a})$

3 Given vector equation $2(\mathbf{x}-\mathbf{a})-3(\mathbf{x}-2\mathbf{b}) = \mathbf{0}$, in terms of $\mathbf{a}$ and $\mathbf{b}$, vector $\mathbf{x}$ can be expressed as (   ).

A. $\frac{6}{5}\mathbf{b}-\frac{2}{5}\mathbf{a}$
B. $\frac{2}{5}\mathbf{a}-\frac{6}{5}\mathbf{b}$
C. $6\mathbf{b}-2\mathbf{a}$
D. $2\mathbf{a}-6\mathbf{b}$

4 In an equilateral triangle ABC with side length 2, the value of $|\overrightarrow{BA}+\overrightarrow{BC}|$ is (   ).

A. 2
B. 4
C. 0
D. $2\sqrt{3}$

B. Fill in the blanks

5 The diagram shows a parallelogram $ABCD$, point E is on side AB and $AB = 3EB$. Given that $\overrightarrow{AB} = \mathbf{a}$ and $\overrightarrow{BC} = \mathbf{b}$, then $\overrightarrow{DE} = $ _____ (in terms of $\mathbf{a}$ and $\mathbf{b}$).

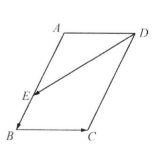

Diagram for question 5

6 The diagram shows a trapezium $ABCD$. Given $AB \parallel CD$, $AB = 3CD$, $\overrightarrow{AB} = \mathbf{a}$ and $\overrightarrow{AD} = \mathbf{b}$, then $\overrightarrow{AO} = $ _____ (in terms of $\mathbf{a}$ and $\mathbf{b}$).

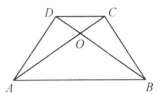

Diagram for question 6

7 In $\triangle ABC$, points D and E are on AB and AC, $DE \parallel BC$, and $\overrightarrow{AD} = 2\overrightarrow{DB}$. Given that $\overrightarrow{AB} = \mathbf{a}$ and $\overrightarrow{AC} = \mathbf{b}$, then, in terms of $\mathbf{a}$ and $\mathbf{b}$, vector $\overrightarrow{DE}$ can be expressed as _____ .

8 The diagram shows a quadrilateral $ABCD$. E, F and G are the midpoints of AD, CD and BD, respectively. Given that $\overrightarrow{AB} = \mathbf{a}$ and $\overrightarrow{BC} = \mathbf{b}$, then $\overrightarrow{EF} = $ _____ (in terms of $\mathbf{a}$ and $\mathbf{b}$).

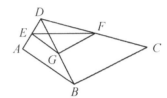

Diagram for question 8

9 The diagram shows $\triangle ABC$, point M is on BC, and $MC = 2BM$. Given that vector $\overrightarrow{AB} = \mathbf{a}$ and $\overrightarrow{AM} = \mathbf{b}$, then vector $\overrightarrow{BC} = $ _____ (in terms of $\mathbf{a}$ and $\mathbf{b}$).

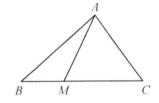

Diagram for question 9

C. Questions that require solutions

10 The diagram shows $\triangle ABC$ with D being the midpoint of AC. Let $\overrightarrow{BD} = \mathbf{a}$ and $\overrightarrow{BC} = \mathbf{b}$. Express vector $\overrightarrow{CA}$ in terms of $\mathbf{a}$ and $\mathbf{b}$.

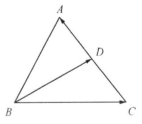

Diagram for question 10

11 The diagram shows a quadrilateral $ABCD$. E and F are the midpoints of AD and BC, respectively. Prove: $\overrightarrow{AB} + \overrightarrow{DC} = 2\overrightarrow{EF}$.

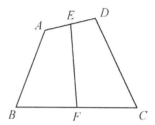

Diagram for question 11

12 In a quadrilateral $ABCD$, $\overrightarrow{AB} = \mathbf{a} + 2\mathbf{b}$, $\overrightarrow{BC} = -4\mathbf{a}$ and $\overrightarrow{CD} = -3\mathbf{a} - 2\mathbf{b}$ ($\mathbf{a}$ and $\mathbf{b}$ are not parallel).

(a) Use linear combination of $\mathbf{a}$ and $\mathbf{b}$ to represent vector $\overrightarrow{AD}$.

(b) Prove that quadrilateral $ABCD$ is a trapezium.

8.10 Linear operations on vectors (2)

Learning objective

Apply linear operations on vectors; use diagrams to represent the results.

A. Multiple choice questions

1 Given non-zero vectors **a**, **b** and **c**, given that $\mathbf{c} = 2\mathbf{a} + \mathbf{b}$, then vector ( ) is parallel to vector **c**.

A. $\mathbf{m} = \mathbf{a} - 2\mathbf{b}$ B. $\mathbf{n} = \mathbf{b} - 2\mathbf{a}$ C. $\mathbf{q} = 4\mathbf{a} + 2\mathbf{b}$ D. $\mathbf{g} = 2\mathbf{a} + 4\mathbf{b}$

2 The diagram shows $\triangle ABC$. D, E and F are the midpoints of AB, AC and BC, respectively. Let $\overrightarrow{BA} = \mathbf{a}$ and $\overrightarrow{BC} = \mathbf{b}$. Then vector ( ) can be expressed as $\frac{1}{2}\mathbf{b} - \mathbf{a}$.

A. $\overrightarrow{AC}$

B. $\overrightarrow{AF}$

C. $\overrightarrow{BE}$

D. $\overrightarrow{CD}$

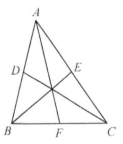

Diagram for question 2

3 Given that, in the rectangle $ABCD$, the diagonals AC and BD intersect at O, $\overrightarrow{AB} = \mathbf{a}$ and $\overrightarrow{BC} = \mathbf{b}$, then $\overrightarrow{CO}$ can be expressed as ( ).

A. $\frac{1}{2}\mathbf{a} + \frac{1}{2}\mathbf{b}$ B. $-\frac{1}{2}\mathbf{a} - \frac{1}{2}\mathbf{b}$ C. $\frac{1}{2}\mathbf{a} - \frac{1}{2}\mathbf{b}$ D. $\frac{1}{2}\mathbf{b} - \frac{1}{2}\mathbf{a}$

4 The diagram shows $\triangle ABC$, with point D on BC. Given that $BD = 2DC$, $\overrightarrow{BA} = \mathbf{a}$ and $\overrightarrow{BC} = \mathbf{b}$, then $\overrightarrow{AD}$ is ( ).

A. $\frac{2}{3}\mathbf{a} - \mathbf{b}$

B. $\frac{2}{3}\mathbf{b} - \mathbf{a}$

C. $\mathbf{b} - \frac{2}{3}\mathbf{a}$

D. $\mathbf{a} - \frac{2}{3}\mathbf{b}$

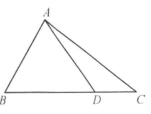

Diagram for question 4

B. Fill in the blanks

5 In $\triangle ABC$, points D, E and F are on AB, AC and BC, $DE \parallel BC$, $DF \parallel AC$, and $\dfrac{AE}{EC} = \dfrac{3}{4}$. Given that $\overrightarrow{BA} = \mathbf{a}$ and $\overrightarrow{BC} = \mathbf{b}$ then, in terms of $\mathbf{a}$ and $\mathbf{b}$, vector $\overrightarrow{FD}$ can be expressed as _____.

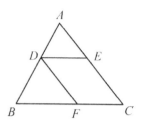

Diagram for question 5

6 Given that $\overrightarrow{AC} = \overrightarrow{AB} + \overrightarrow{AD}$, where $\overrightarrow{AC} = \mathbf{a}$ and $\overrightarrow{BD} = \mathbf{b}$, then $\overrightarrow{AB} =$ _____ , $\overrightarrow{AD} =$ _____ .

7 The diagram shows a trapezium $ABCD$ with $AD \parallel BC$. Points E and F are the midpoints of AB and DC, respectively.

Given that $\overrightarrow{AD} = \mathbf{a}$ and $\overrightarrow{EF} = \mathbf{b}$, then $\overrightarrow{BC} =$ _____ (in terms of $\mathbf{a}$ and $\mathbf{b}$).

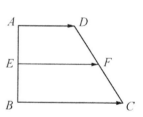

Diagram for question 7

8 In the diagram, points D and E are the midpoints of AC and AB, respectively. Let $\overrightarrow{BO} = \mathbf{a}$ and $\overrightarrow{OC} = \mathbf{b}$.

Then $\overrightarrow{ED} =$ _____ (in terms of $\mathbf{a}$ and $\mathbf{b}$).

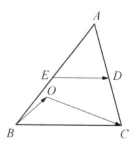

Diagram for question 8

C. Questions that require solutions

9 The diagram shows a parallelogram $ABCD$. The diagonals AC and BD intersect at point O. E is a point on BC such that $EC = \dfrac{1}{4}BC$, DE and AC intersect at point F. Given that $\overrightarrow{BA} = \mathbf{a}$ and $\overrightarrow{BC} = \mathbf{b}$, express vectors $\overrightarrow{FD}$ and $\overrightarrow{FC}$ in terms of $\mathbf{a}$ and $\mathbf{b}$.

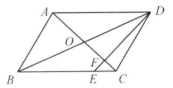

Diagram for question 9

10 In parallelogram $ABCD$, point E is on AD and $AE = 3ED$. CE is extended beyond E to point F such that $EF = CE$. Let $\overrightarrow{BA} = \mathbf{a}$ and $\overrightarrow{BC} = \mathbf{b}$. Express vectors $\overrightarrow{CE}$ and $\overrightarrow{AF}$ in terms of $\mathbf{a}$ and $\mathbf{b}$.

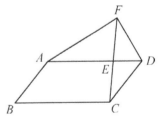

Diagram for question 10

8.11 Column vectors and their operations (1)

Learning objective

Understand the concepts of column vectors and apply linear operations on them.

A. Multiple choice questions

1. The diagram shows a vector on a coordinate plane. The incorrect representation of the vector is (   ).

   A. $\overrightarrow{OA}$

   B. $\begin{pmatrix} 3 \\ 5 \end{pmatrix}$

   C. $\mathbf{a}$

   D. $\begin{pmatrix} 5 \\ 3 \end{pmatrix}$

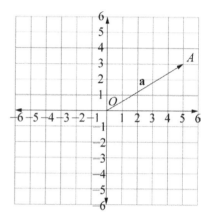

Diagram for question 1

2. Given $\mathbf{a} = \begin{pmatrix} -1 \\ 3 \end{pmatrix}$ and the points $A(-1, 3)$ and $B(2, -6)$, then (   ) of the equations is/are correct.

   ① $\mathbf{a} = \overrightarrow{OA}$ ② $\overrightarrow{OB} = -2\mathbf{a}$ ③ $\overrightarrow{AB} = -3\mathbf{a}$ ④ $\overrightarrow{OA} + \overrightarrow{OB} + \mathbf{a} = \mathbf{0}$

   A. one   B. two   C. three   D. four

3. Given $P(-1, 0)$, $Q(3, 0)$ and $M(0, 1)$, any of these points, each denoted by N but with different coordinates, **except** (   ) could be the fourth point in a quadrilateral $PQMN$.

   A. $N(2, -1)$ B. $N(4, 1)$ C. $N(-4, 1)$ D. $N(1, 2)$

B. Fill in the blanks

4. Given column vectors $\mathbf{a} = \begin{pmatrix} 3 \\ 2 \end{pmatrix}$ and $\mathbf{b} = \begin{pmatrix} -1 \\ -3 \end{pmatrix}$, then $2\mathbf{a} + 3\mathbf{b} = $ _____.

> When a vector is expressed **in column form** $\begin{pmatrix} x \\ y \end{pmatrix}$, where x and y are real numbers, it is also known as **a column vector**.

5 Given vectors $\mathbf{a} = \begin{pmatrix} 2 \\ -3 \end{pmatrix}$ and $\mathbf{b} = \begin{pmatrix} 1 \\ -2 \end{pmatrix}$, then $|2\mathbf{a} - \mathbf{b}| = $ _____.

6 Translating column vector $\mathbf{a} = \begin{pmatrix} 2 \\ 1 \end{pmatrix}$ 2 units to the right and 3 units downwards gives the resultant vector _____.

7 Given vectors $\overrightarrow{AB} = \begin{pmatrix} 2 \\ -3 \end{pmatrix}$ and $\overrightarrow{BC} = \begin{pmatrix} 1 \\ 2 \end{pmatrix}$, then $|\overrightarrow{AC}| = $ _____.

8 In the parallelogram $ABCD$, with vertices $A(2, -1)$, $B(3, 1)$ and $C(0, 3)$, then $\overrightarrow{CD} = $ _____ in column form.

9 Given two points $A(1, 2)$ and $B(3, 1)$, with point C on the x-axis, and with A, B and C being collinear (on the same line), then the coordinates of point C are _____.

C. Questions that require solutions

10 Given vectors $\mathbf{a} = \begin{pmatrix} 2 \\ -1 \end{pmatrix}$, $\mathbf{b} = \begin{pmatrix} 1 \\ 2 \end{pmatrix}$ and $\mathbf{c} = \begin{pmatrix} 3 \\ -4 \end{pmatrix}$, find:
 (a) $2\mathbf{a} + 3\mathbf{b} - 3\mathbf{c}$ (b) $|3\mathbf{a} - 2\mathbf{b}|$.

11 Given vectors $\mathbf{a} + 2\mathbf{b} = \begin{pmatrix} 3 \\ -1 \end{pmatrix}$, $\mathbf{a} - \mathbf{b} = \begin{pmatrix} 0 \\ 2 \end{pmatrix}$, and $\mathbf{c} = \begin{pmatrix} 3 \\ -4 \end{pmatrix}$, find:
 (a) vectors $\mathbf{a}$ and $\mathbf{b}$ (b) $|3\mathbf{a} - 2\mathbf{c}|$.

12 On a coordinate plane, $A(2, 2)$, $B(-3, 4)$ and $C(-1, 2)$ are three vertices of parallelogram $ABCD$.

(a) Find the coordinate of point D.

(b) Find the length of diagonal BD.

(c) Given that the diagonals of parallelogram $ABCD$ intersect at point G, find the column vector $\overrightarrow{AG}$.

8.12 Column vectors and their operations (2)

Learning objective

Solve problems involving column vectors; use vectors to construct geometric arguments and proofs.

A. Multiple choice questions

1 Given $\mathbf{a} = \begin{pmatrix} 2 \\ -3 \end{pmatrix}$ and $\mathbf{b} = \begin{pmatrix} -1 \\ 1 \end{pmatrix}$, the unit vector in

the same direction as $\mathbf{a} - \mathbf{b}$ is (　　).

A. $\begin{pmatrix} -\dfrac{3}{5} \\ \dfrac{4}{5} \end{pmatrix}$　　　　B. $\begin{pmatrix} \dfrac{3}{5} \\ -\dfrac{4}{5} \end{pmatrix}$

C. $\begin{pmatrix} \dfrac{3}{5} \\ \dfrac{4}{5} \end{pmatrix}$　　　　D. $\begin{pmatrix} -\dfrac{3}{5} \\ -\dfrac{4}{5} \end{pmatrix}$

> Note: **A unit vector** is a vector with magnitude 1, in any direction. In general, $\dfrac{\mathbf{a}}{|\mathbf{a}|}$ is a unit vector in the same direction as $\mathbf{a}$.

2 Given vector $\mathbf{a} = \begin{pmatrix} 1 \\ 3 \end{pmatrix}$, $|\mathbf{b}| = 2\sqrt{10}$, and $\mathbf{a} /\!/ \mathbf{b}$, then vector $\mathbf{b}$ is (　　).

A. $\begin{pmatrix} -2 \\ 6 \end{pmatrix}$

B. $\begin{pmatrix} 2 \\ 6 \end{pmatrix}$

C. $\begin{pmatrix} 6 \\ -2 \end{pmatrix}$ or $\begin{pmatrix} -6 \\ 2 \end{pmatrix}$

D. $\begin{pmatrix} -2 \\ -6 \end{pmatrix}$ or $\begin{pmatrix} 2 \\ 6 \end{pmatrix}$

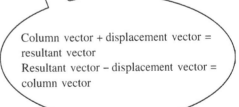

> Column vector + displacement vector = resultant vector
> Resultant vector – displacement vector = column vector

3 A column vector has been translated 5 units to the left and 3 units upwards. The displacement vector is (　　).

A. $\begin{pmatrix} 5 \\ 3 \end{pmatrix}$　　　　　　　　B. $\begin{pmatrix} 5 \\ -3 \end{pmatrix}$

C. $\begin{pmatrix} -5 \\ 3 \end{pmatrix}$　　　　　　　　D. $\begin{pmatrix} -5 \\ -3 \end{pmatrix}$

4 In question 3, if the starting column vector is $\begin{pmatrix} 2 \\ -3 \end{pmatrix}$, then after the translation, the resultant vector is (   ).

A. $\begin{pmatrix} 5 \\ 3 \end{pmatrix}$ 　　　 B. $\begin{pmatrix} 2 \\ -3 \end{pmatrix}$ 　　　 C. $\begin{pmatrix} 7 \\ 0 \end{pmatrix}$ 　　　 D. $\begin{pmatrix} -3 \\ 0 \end{pmatrix}$

5 Look at the coordinate plane shown on the right. Of these statements, (   ) is incorrect.

A. $\overrightarrow{AB}$ and $\overrightarrow{CD}$ are parallel.

B. $|\overrightarrow{AB}| = 2|\overrightarrow{CD}|$

C. $\overrightarrow{CA} + \overrightarrow{AB} = \overrightarrow{CD} + \overrightarrow{DB}$

D. $\overrightarrow{CA} - \overrightarrow{AB} = \overrightarrow{CD} - \overrightarrow{DB}$

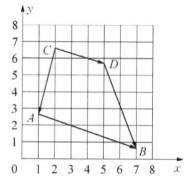

Diagram for question 5

B. Fill in blanks

6 Given non-zero vector $\mathbf{a} = \begin{pmatrix} x_1 \\ y_1 \end{pmatrix}$ and $\mathbf{b} = \begin{pmatrix} x_2 \\ y_2 \end{pmatrix}$, then $|\mathbf{a}| = $ _____, $|\mathbf{b}| = $ _____ and $|\mathbf{a} - \mathbf{b}| = $ _____.

7 Given vectors $\mathbf{a} = \begin{pmatrix} 1 \\ 3 \end{pmatrix}$, $\mathbf{b} = \begin{pmatrix} -2 \\ -6 \end{pmatrix}$, $\mathbf{c} = \begin{pmatrix} 3 \\ -1 \end{pmatrix}$ and $\mathbf{d} = \mathbf{b} - \mathbf{a}$, the vectors that are parallel to each other are _____.

8 Given vector $\mathbf{a} = \begin{pmatrix} 1 \\ 2 \end{pmatrix}$, the vector parallel to $\mathbf{a}$ with magnitude 5 is _____.

9 Given vectors $\mathbf{a} = \begin{pmatrix} 3 \\ x \end{pmatrix}$ and $\mathbf{b} = \begin{pmatrix} 2x - 1 \\ 1 \end{pmatrix}$, if $\mathbf{a} \,/\!/\, \mathbf{b}$, then $x = $ _____.

10 Given vectors $\mathbf{a} = \begin{pmatrix} 2 \\ 1 \end{pmatrix}$, $\mathbf{b} = \begin{pmatrix} 1 \\ x \end{pmatrix}$ such that $\mathbf{a} - 3\mathbf{b}$ and $2\mathbf{a} + \mathbf{b}$ are parallel, then $x = $ _____.

C. Questions that require solutions

11 $A(1, 1)$, $B(2, 2)$ and $C(1, 3)$ are three vertices of a rhombus $ABCD$.

(a) Find the coordinates of the fourth vertex D, and draw quadrilateral $ABCD$ on the coordinate plane supplied.

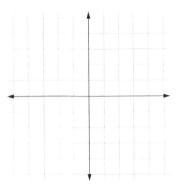

(b) Use a column vector to describe a translation that translates CB to DA.

Diagram for question 11

(c) Find the area of this rhombus.

12 $\overrightarrow{OP} = \begin{pmatrix} 1 \\ 2x \end{pmatrix}$ and $\overrightarrow{OQ} = \begin{pmatrix} 2 \\ x + 1 \end{pmatrix}$. A parallelogram is formed with points O, P, Q and a fourth point A, such that the value of $|\overrightarrow{PQ}|$ is a minimum.

(1) Find an expression for $\overrightarrow{OA}$.

(2) Find the area of the triangle with all the point As that satisfy the condition.

13 The diagram shows a trapezium $ABCD$ with $AB \parallel DC$. Points E and F are the midpoints of AD and BC, respectively. Use column vectors to prove:

(a) $EF \parallel AB \parallel DC$

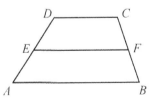

(b) $EF = \dfrac{1}{2}(AB + DC)$.

Diagram for question 13

(Hint: you may establish a coordinate plane with A at the origin and AB as the x-axis, and then use column vectors to prove this result.)

Unit test 8

A. Multiple choice questions

1. Of these equations, the incorrect one is (   ).

   A. $\mathbf{0} - \mathbf{0} = \mathbf{0}$

   B. $\mathbf{a} - \mathbf{a} = 0$

   C. $\mathbf{a} - \mathbf{0} = \mathbf{a}$

   D. $(\mathbf{a} + \mathbf{b}) - \mathbf{c} = (\mathbf{a} - \mathbf{c}) + \mathbf{b}$

2. Of these statements, (   ) is incorrect.

   A. $0\mathbf{a} = \mathbf{0}$

   B. If $\mathbf{a} = \dfrac{1}{2}\mathbf{b}$ ($\mathbf{b}$ is non-zero vector), then $\mathbf{a} /\!/ \mathbf{b}$.

   C. Let $\mathbf{e}$ be a unit vector, then $|\mathbf{e}| = 1$.

   D. If $|\mathbf{a}| = |\mathbf{b}|$, then $\mathbf{a} = \mathbf{b}$ or $\mathbf{a} = -\mathbf{b}$

3. If $\mathbf{a} + \mathbf{b} = 2\mathbf{c}$, $\mathbf{a} - \mathbf{b} = 3\mathbf{c}$ and $\mathbf{c} \neq \mathbf{0}$, then $\mathbf{a}$ and $\mathbf{b}$ are (   ).

   A. equal vectors

   B. parallel vectors

   C. acting in the same direction, but have different lengths

   D. acting in different directions, but have the same length

4. Look at the coordinate plane. The incorrect statement is (   ).

   A. $\overrightarrow{MN}$ and $\overrightarrow{PQ}$ are equal.

   B. $\overrightarrow{MN}$ and $\overrightarrow{PQ}$ are parallel.

   C. $\overrightarrow{MN}$ and $\overrightarrow{PQ}$ are in the same direction.

   D. $\overrightarrow{MN}$ and $\overrightarrow{PQ}$ are in opposite directions.

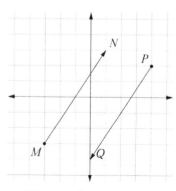

Diagram for question 4

B. Fill in the blanks

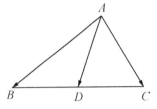

5. The diagram shows $\triangle ABC$, with point D on BC.
   Use vectors $\overrightarrow{AB}$, $\overrightarrow{AC}$ and $\overrightarrow{AD}$ to represent
   vector $\overrightarrow{BD}$ = _____ and vector $\overrightarrow{DC}$ = _____ .

Diagram for question 5

6. The diagram shows $\triangle ABC$, in which D and E are the midpoints of BC and AC, respectively. AD intersects BE at point G. Let $\overrightarrow{AB}$ = **a** and $\overrightarrow{AD}$ = **b**. Then $\overrightarrow{BE}$ = _____ (in terms of **a** and **b**).

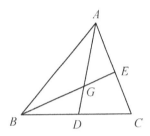

Diagram for question 6

7. After column vector $\mathbf{a} = \begin{pmatrix} -3 \\ 4 \end{pmatrix}$ is translated 9 units to the right and 2 units upwards, the resultant vector is _____ and the magnitude is _____ .

8. Given points $A(4, 6)$, $B(-3, 2)$, and $\overrightarrow{AP} = -\dfrac{2}{3} \overrightarrow{PB}$, the the coordinates of P are _____ .

9. Given vectors $\overrightarrow{OA} = \begin{pmatrix} 2 \\ -1 \end{pmatrix}$ and $\overrightarrow{AB} = \begin{pmatrix} 1 \\ 2 \end{pmatrix}$, then $|\overrightarrow{OB}|$ = _____ .

10. Given $\overrightarrow{OP_1} = \begin{pmatrix} -1 \\ -6 \end{pmatrix}$, $\overrightarrow{OP_2} = \begin{pmatrix} 3 \\ 0 \end{pmatrix}$, point P is on the extension of $P_2 P_1$ beyond P_1, and $|\overrightarrow{PP_1}| = \dfrac{1}{3} |\overrightarrow{P_1 P_2}|$, then $\overrightarrow{OP}$ = _____ (in column form).

C. Questions that require solutions

11. The diagram shows vectors **a**, **b** and **c**. Construct these vectors.
    (Just draw them, there is no need to write the steps.)
    (a) $\mathbf{a} + \mathbf{b}$ (b) $\mathbf{b} - \mathbf{c}$ (c) $\mathbf{a} - (\mathbf{b} + \mathbf{c})$

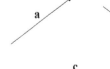

Diagram for question 11

12 The diagram shows a parallelogram $ABCD$, with diagonals AC and BD intersecting at point O. Let $\overrightarrow{OA}$ = **a** and $\overrightarrow{OB}$ = **b**. Express these vectors in terms of **a** and **b**.

(a) $\overrightarrow{OC}$ (b) $\overrightarrow{OD}$ (c) $\overrightarrow{AB}$ (d) $\overrightarrow{BC}$

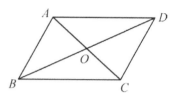

Diagram for question 12

13 Given that $\mathbf{a} = \begin{pmatrix} 2 \\ -3 \end{pmatrix}$ and $\mathbf{b} = \begin{pmatrix} -3 \\ 2 \end{pmatrix}$, find the vector that is parallel to **a** + **b** and has a magnitude of 2. Express the vector in column form.

14 There four points $A(-2, -2)$, $B(0, 3)$, $C(3, 3)$ and $D(1, -2)$ lie on a coordinate plane, as shown in the diagram.

(a) Draw these points on the coordinate plane and then join AB, BC, CD and BA to obtain a quadrilateral $ABCD$.

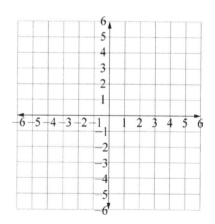

Diagram for question 14

(b) Use a column vector to describe the translation that translates point A to point C.

(c) Use column vectors to prove that quadrilateral $ABCD$ is a parallelogram.

Chapter 9　Circles and properties of circles

9.1　Arc length

Learning objective

Understand the concept of arc length and calculate arc lengths in circles.

A. Multiple choice questions

1 To calculate the length of an arc of a circle, we need to know (　　).

A. the diameter
B. the radius
C. the central angle
D. the radius and the central angle

2 (　　) of these statements is/are correct.

① The greater the radius, the longer is the arc length.

② If two arcs are of the same length, then their central angles are equal.

③ If the central angle is increased to 3 times the original size and the radius is decreased to $\frac{1}{3}$ of the original length, then the arc length remains unchanged.

④ A chord divides the circumference of a circle into two parts. If one part is a minor arc, then the other part is a major arc.

A. None　　　　B. One　　　　C. Two　　　　D. Three

3 If the central angle of an arc is $n°$, then the incorrect expression for calculating the arc length l is (　　). (Note: r = radius; d = diameter; C = circumference)

A. $l = \frac{n}{360}\pi r$　　　B. $l = \frac{n}{180}\pi r$　　　C. $l = \frac{n}{360}\pi d$　　　D. $l = \frac{n}{360}C$

B. Fill in the blanks

4 In a circle, if the length of an arc is $\frac{1}{5}$ of the circumference, then the central angle of the arc is _____.

5 The radius of an arc of a circle is 6 cm and its central angle is 60°. The arc length is _____ cm.

6 In the diagram, the arc length is 18.84 cm. Then the radius of the circle of which the arc is a part is _____ cm.

Diagram for question 6

7 As shown in the diagram, arcs are drawn using the vertices of the two acute angles in a set square of 45° as the centres of circles and the lengths of the non-hypotenuse sides as the radii. Given that points B, D, C and E are all on the same line, then the arc length $\overarc{AE}$ is _____ times the arc length $\overarc{AD}$.

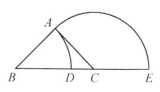

Diagram for question 7

8 A cylindrical log of radius 5 cm was sawn into 10 equal circular sectors. The perimeter of each circular sector is _____ cm.

C. Questions that require solutions

9 Fill in the table.

Radius	Central angle	Arc length
12 cm	45°	
	60°	8.37 cm
1.9 m		0.994 m

10 In the diagram, C and D are the two points on the diameter AB of a semicircle, and AB = 20 metres. Three smaller semicircles have AC, CD and DB as their diameters. Ming and Larry started walking from point A towards point B at the same time. Ming walked along the larger semicircle while Larry walked along the 3 smaller semicircles. How many metres did Ming and Larry walk, respectively?

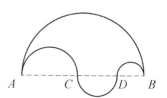

Diagram for question 10

11 Given that the length of an arc with central angle 120° is 62.8 cm, what is the diameter of the circle of which the arc is part?

12 In the diagram, the quadrilateral *ABCD* is a rectangle with length 10 cm and width 6 cm. Find the perimeter of the shaded region.

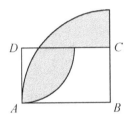

Diagram for question 12

13 Find the perimeter of the shaded region of this diagram.

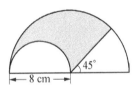

Diagram for question 13

9.2 The area of a sector

Learning objective

Calculate the area of a sector of a circle.

A. Multiple choice questions

1. The radius of a sector is 5 cm and its central angle is 36°. The area of the sector is ( ).

A. 3.14 cm²   B. 3.925 cm²   C. 7.85 cm²   D. 15.7 cm²

2. The arc length of a sector is 10 cm and the radius is 20 cm. The area of the sector is ( ).

A. 62.4 cm²   B. 100 cm²   C. 200 cm²   D. 314 cm²

3. Given that the area of a sector is S and the area of the circle containing the sector is $5S$, then the central angle of the sector is ( ).

A. 5°   B. 10°   C. 36°   D. 72°

B. Fill in the blanks

4. Fill in the table.

Radius, r	Central angle, n	Arc length	Area of sector, S
12 mm		56.52 mm	
2 cm	72°		
	60°		13.083 cm²
	18°	π m	

5. The minute hand on a clock is 6 cm long. From 2:05 to 2:25 in the afternoon, the area that the minute hand sweeps out is _____ cm².

6. Given that the circumference of a circle is 12.56 cm, then the area of the sector with central angle 108° is _____ cm².

7 The area of a sector of a circle, with central angle 18°, is 36 cm². The area of the whole circle is _____ cm².

8 Given that the central angles of two sectors are equal and the radius of one of the sectors is $\frac{1}{5}$ of the radius of the other, then the area of the larger sector is _____ times the area of the smaller sector.

C. Questions that require solutions

9 In the diagram, the length of the rectangle is 16 cm and its width is 10 cm. Find the area of the shaded region.

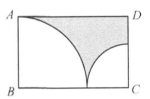

Diagram for question 9

10 In the diagram, the length of each non-hypotenuse side of the right-angled isosceles triangle is 6 cm. Find the area of the shaded region.

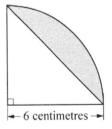

6 centimetres

Diagram for question 10

11 In the diagram, the radius of the circle $r = 4$ cm and the quadrilateral $ABCD$ is a parallelogram. Find the area of the shaded region.

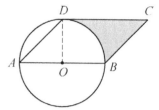

Diagram for question 11

12 In the diagram, the length of the side of the square $ABCD$ is 20 cm. A semicircle is drawn with AB as diameter. Find the area of the shaded region.

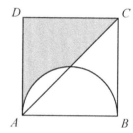

Diagram for question 12

13 In the diagram, AB is the diameter of the semicircle with centre O. $AB = 20$ cm and C is the midpoint of arc AB. ABD is a sector. Find the area of the shaded region.

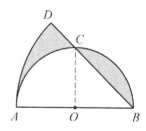

Diagram for question 13

9.3 Determining a circle

Learning objective

Identify and apply circle definitions and properties to determine a circle.

A. Multiple choice questions

1. The radius of a circle, centre O, is 4 cm, and A is the midpoint of OP. Given that $OP = 7$ cm, then (   ).

   A. point A is inside the circle
   B. point A is on the circle
   C. point A is outside the circle
   D. uncertain

2. From a point to a circle, the longest distance is 11 and the shortest distance is 5. The radius of the circle is (   ).

   A. 16 or 6
   B. 3 or 8
   C. 3
   D. 8

3. Read these statements.
   ① Infinitely many circles can be constructed passing through one point.
   ② Infinitely many circles can be constructed passing through two points.
   ③ Only one circle can be constructed passing through three points.
   ④ There might be no circle that can be drawn passing through four points.
   (   ) of the above statements is/are correct.

   A. One
   B. Two
   C. Three
   D. Four

B. Fill in the blanks

4. Let the radius of a circle be R and the distance from point P to the centre be d. Then:
   (a) point P is outside the circle $\Leftrightarrow d$ _____ R
   (b) point P is on the circle $\Leftrightarrow d$ _____ R
   (c) point P is inside the circle $\Leftrightarrow d$ _____ R.

5. The radius of a circle, centre O, is 4 cm and A is the midpoint of line segment OP.
   (a) When $OP = 6$ cm, point A is _____ the circle.
   (b) When $OP = 8$ cm, point A is _____ the circle.
   (c) When $OP = 10$ cm, point A is _____ the circle.

6 There is/are _____ circle(s) passing through two points on a plane, and the centres of the circles must be on the _____ of the line segment connecting the two points.

7 There is/are only _____ circle(s) passing through three non-collinear points (points not all on the same line) in a plane, and the centre of the circles is the point of intersection of the _____ of the three sides of the triangle with the three points as vertices.

8 In a coordinate plane, the circle with centre (a, b) and radius r is the set of all the points (x, y) that satisfy the equation: _____.

C. Questions that require solutions

9 In a coordinate plane, the radius of a circle is 8 and the coordinates of its centre are $(-1, 5)$. Explain the relationship between the positions of point $P(3, -2)$ and the circle.

10 Given a linear function $y = x + 1$ and a circle with centre $(2, 3)$ and radius $3\sqrt{2}$, find the coordinates of the points of intersection of the graph of the function and the circle. Draw the line and the circle on the coordinate plane and indicate the intersecting points.

(Take $\sqrt{2} \approx 1.4$)

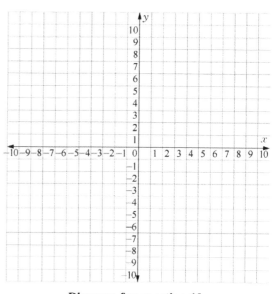

Diagram for question 10

11 The diagram shows an isosceles trapezium $ABCD$, $AD \parallel BC$, $AD = 3$, $AB = CD = 4$ and $\angle ABC = 60°$. The bisector of $\angle ABC$ intersects DC at point E and the extension of AD beyond D at point F. P is a moving point on BE (including points B and E). Now draw a circle with point P as centre and BP as radius. What is the range of the length of BP when point A is inside the circle and point E is outside the circle?

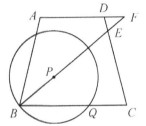

Diagram for question 11

9.4 Central angle, arc, chord, distance from chord to centre and their relationships (1)

Learning objective

Understand and apply the relationships between a circle's central angles, arcs, chords and the distances from chords to the centre of the circle.

A. Multiple choice questions

1 Read these statements.

① A diameter is a chord.

② A chord is a diameter.

③ A semicircle is an arc, but an arc is not necessarily a semicircle.

④ Two arcs with equal lengths are equal arcs.

(   ) of the statements is/are correct.

A. One   B. Two   C. Three   D. Four

2 Read these statements. The correct one is (   ).

A. A chord is a diameter.

B. A semicircle is an arc.

C. A line segment passing through the centre of a circle is a diameter.

D. Two circles with the same centre and equal radius are concentric circles.

3 In a circle, centre O, if the central angle $\angle BOA$ is twice the central angle $\angle COD$, then (   ) is true.

A. $AB = 2CD$

B. $\overset{\frown}{AB} = 2\,\overset{\frown}{CD}$

C. $\overset{\frown}{AB} = \overset{\frown}{CD}$

D. $AB = CD$

4 In a circle, centre O, if the central angle $\angle AOB = 90°$ and the distance from O to chord AB is 4. Then the diameter of the circle is (   ).

A. $4\sqrt{2}$

B. $8\sqrt{2}$

C. 24

D. 16

B. Fill in the blanks

5 The section between any two points on a circle is called _____. The line segment that joins any two points on a circle is called _____.

6 In the diagram, AD is the diameter of the circle, centre O, and $\angle AOB = \angle BOC = \angle COD$. The central angle of $\overset{\frown}{AB}$ is _____ degrees, the arcs equal to $\overset{\frown}{AB}$ are _____, and BD is _____ to CO.

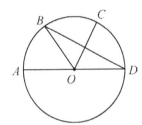

Diagram for question 6

7 In a circle, the angle subtended by an arc at the centre is _____ the angle subtended by the same arc at the circumference.

8 In the diagram, AB and CD are two diameters of a circle, centre O. Chord $CE \parallel AB$ and $\angle EOC = 40°$. $\angle BOC =$ _____.

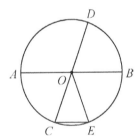

Diagram for question 8

9 In the diagram, CD is a diameter of a circle, centre O, and E is a point on the circle. $\angle EOD = 45°$. A is a point on the extension of DC beyond C, and AE intersects the circle at point B. Given that $AB = OC$, then $\angle EAD =$ _____.

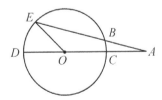

Diagram for question 9

C. Questions that require solutions

10 In the diagram, AB is a chord of a circle, centre O, $AC = BD$ and radii OE and OF pass through points C and D. Prove that $\overset{\frown}{AE} = \overset{\frown}{BF}$.

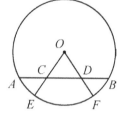

Diagram for question 10

11 In the diagram, OA and OB are radii of a circle, centre O. C is on $\overset{\frown}{AB}$, $CD \perp OA$ at D, $CE \perp OB$ at E and $CD = CE$. Prove that $\overset{\frown}{AC} = \overset{\frown}{BC}$.

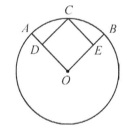

Diagram for question 11

12 In the diagram, AB is a diameter of a circle, centre O, and $CO \perp AB$. Point D is the midpoint of CO and $DE \mathbin{/\mkern-5mu/} AB$. Prove that $\overset{\frown}{CE} = 2\,\overset{\frown}{AE}$.

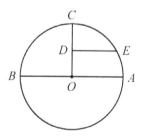

Diagram for question 12

9.5 Central angle, arc, chord, distance from chord to centre and their relationships (2)

 Learning objective

Understand and apply the relationships between a circle's central angles, arcs, chords and the distances from chords to the centre of the circle.

 A. Multiple choice questions

1. In the same circle, if two chords are equal, statement (   ) is incorrect.
   A. The major arcs subtended by the two chords are equal.
   B. The central angles subtended by the two chords are equal.
   C. The distances from both chords to the centre are equal.
   D. The two chords must be parallel.

2. Given that $\overarc{AB}$ and $\overarc{CD}$ are two arcs of the same circle, and $\overarc{AB} = 2\overarc{CD}$, then the relation between the chords AB and CD is (   ).
   A. $AB = 2CD$ B. $AB < 2CD$ C. $AB > 2CD$ D. uncertain

 B. Fill in the blanks

3. In the diagram, AB and CD are two chords of a circle, centre O, OE and OF are the distances from chords AB and CD to the centre O, respectively.
   (a) If $AB = CD$, then _____ , _____ and _____ .
   (b) If $OE = OF$, then _____ , _____ and _____ .
   (c) If $\overarc{AB} = \overarc{CD}$, then _____ , _____ and _____ .
   (d) If $\angle AOB = \angle COD$, then _____ , _____ and _____ .

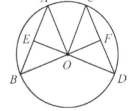

Diagram for question 3

4. In the diagram, AB and DE are diameters of a circle, centre O, and chords $AC \parallel DE$. Given that $BE = 3$, then $CE =$ _____ .

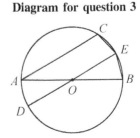

Diagram for question 4

5 The diagram shows a circle, centre O. Given that $\angle BAC = 40°$, then $\angle BDC$ _____ and $\angle BOC$ = _____.

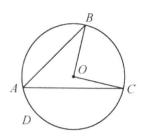

Diagram for question 5

6 $\triangle ABC$ is a triangle inscribed in a circle, centre O. Given that $\overset{\frown}{AB} = \overset{\frown}{BC}$ and $\angle B = 50°$, then $\angle A$ = _____.

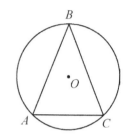

Diagram for question 6

7 In the diagram, the quadrilateral $ABCD$ is a rhombus with side length 2 cm, points E, B, C and F are all on the arc of a circle with centre D, and $\angle ADE = \angle CDF$. The length of $\overset{\frown}{EF}$ is _____ cm. (Leave your answer in terms of π.)

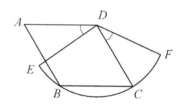

Diagram for question 7

8 In the diagram, points A, B and C are on a circle, centre O, with radius 2 cm. Quadrilateral $OABC$ is a rhombus. Then the area formed by arc $\overset{\frown}{BC}$ and chord BC is _____ cm^2.

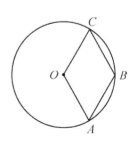

Diagram for question 8

C. Questions that require solutions

9 The diagram shows a circle, centre O, with $\overset{\frown}{AC} = \overset{\frown}{CB}$. D and E are the midpoints of the radii OA and OB, respectively. Prove that $CD = CE$.

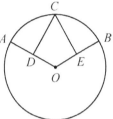

Diagram for question 9

10 In the diagram, AD and BC are two chords of a circle, centre O, and $AD = BC$. Prove that $AB = CD$.

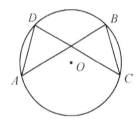

Diagram for question 10

11 The diagram shows a sector OAD with $\angle AOD = 90°$. Given that $AB = BC = CD$, and AC and BD intersect at point K, find the size of $\angle AKD$.

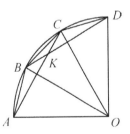

Diagram for question 11

9.6 Perpendicular chord bisector theorem (1)

Learning objective

Understand and apply the perpendicular chord bisector theorem.

A. Multiple choice questions

1 Read these three statements.

① A circle has both line symmetry and rotational symmetry.

② A diameter perpendicular to a chord bisects the chord.

③ The arcs subtended by equal central angles are equal.

The true statements are (   ).

A. ① and ②   B. ② and ③

C. ① and ③   D. all the three

2 The diagram shows a circle, centre O, with $MN \perp AB$ at C. Equation (   ) is incorrect.

A. $AC = BC$

B. $\overgroup{AN} = \overgroup{BN}$

C. $\overgroup{AM} = \overgroup{BM}$

D. $OC = CN$

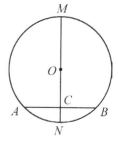

Diagram for question 2

3 In the diagram, the diameter of a circle, centre O, is 10, the length of chord AB is 6, M is a moving point on chord AB. Then the set of values that the segment OM can take is (   ).

A. $3 \leqslant OM \leqslant 5$

B. $4 \leqslant OM \leqslant 5$

C. $3 < OM < 5$

D. $4 < OM < 5$

Diagram for question 3

B. Fill in the blanks

4 In a circle with radius 13, the length of chord AB is 24. The distance from chord AB to the centre is _____.

5 In the diagram, the radius of a circle, centre O, is 10, $OC \perp AB$ at D, $AB = 16$. The length of CD is _____.

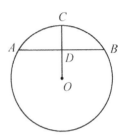

Diagram for question 5

6 A circle, centre O, has a diameter of 10 cm. $\triangle ABC$ is an inscribed isosceles triangle in the circle, with $AB = AC$, and $BC = 6$ cm. Then the area of $\triangle ABC$ is _____.

7 The radius of a circle, centre O, is 5, chord $AB \parallel CD$ and $AB = 8$. If the distance between AB and CD is 1, then the length of CD is _____.

C. Questions that require solutions

8 The diagram shows a circle, centre O. Line segment AB intersect the circle at points C and D, and $OA = OB$. Prove that $AC = BD$.

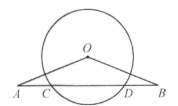

Diagram for question 8

9 The diagram shows the cross-section of a water pipe with diameter 100 cm. If the width of the water inside the pipe is $AB = 60\,\text{cm}$, then what is the maximu depth of the water?

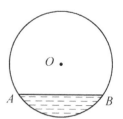

Diagram for question 9

10 The diagram shows a circle, centre O, with radius 2. Point O is on BC and the circle intersects BA at points E and F. Given that $\sin \angle ABC = \dfrac{1}{3}$ and $EF = 2\sqrt{3}$, find the length of BO.

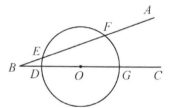

Diagram for question 10

11 The digram shows a coordinate plane, with origin O, and point A is on the negative side of the y-axis. A is the centre of a circle, with radius 5, intersecting the x-axis at B and C and the y-axis at D and E. Given that $\tan \angle DBO = \dfrac{1}{2}$, find the coordinates of D.

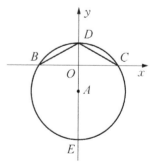

Diagram for question 11

9.7 Perpendicular chord bisector theorem (2)

 Learning objective

Understand and apply the perpendicular chord bisector theorem.

 A. Multiple choice questions

1 Read these statements:

① A line that is perpendicular to and bisects a chord passes through the centre of the circle.

② A diameter that is perpendicular to a chord bisects its arc.

③ A diameter that bisects a chord must be perpendicular to the chord.

④ A diameter that is perpendicular to a chord must bisect the chord.

(   ) of the statements is/are correct.

  A. One   B. Two   C. Three   D. Four

2 In the diagram, AB is a diameter of a circle, centre O, and $CD \perp AB$ at E. If $AB = 10$ and $CD = 8$, then the length of AE is (   ).

  A. 2

  B. 3

  C. 4

  D. 5

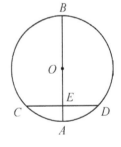

Diagram for question 2

3 In the diagram, AB is a chord of a circle, centre O, and $OC \perp AB$ at D. To make quadrilateral $OACB$ a rhombus, one more condition is needed. The condition can be (   ).

  A. $AD = BD$

  B. $OD = CD$

  C. $\angle CAD = \angle CBD$

  D. $\angle OCA = \angle OCB$

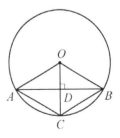

Diagram for question 3

4 Given that AB is a chord of a circle, centre O, with radius r, and $AB = r$, then $\angle AOB =$ (   ).

  A. 30°   B. 60°   C. 90°   D. not sure

B. Fill in the blanks

⑤ The diagram shows a circle, centre O. Chord AB intersects radius OC at point D.

(a) If $\overset{\frown}{AC} = \overset{\frown}{BC}$, then:

$$\angle \underline{\hspace{1.5cm}} = \angle \underline{\hspace{1.5cm}}$$
$$OC \underline{\hspace{2cm}} AB$$
$$AD \underline{\hspace{2cm}} BD.$$

(b) If $AD = BD$, then:

$$\overset{\frown}{AC} \underline{\hspace{2cm}} \overset{\frown}{BC}$$
$$\angle \underline{\hspace{1.5cm}} = \angle \underline{\hspace{1.5cm}}$$
$$OC \underline{\hspace{2cm}} AB.$$

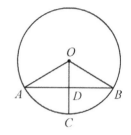

Diagram for question 5

⑥ In the diagram, AB is a diameter of a circle, centre O, chord $CE \perp AB$ at D. Given that $AB = 4$ and $AC = 2\sqrt{3}$, then $CE = \underline{\hspace{2cm}}$.

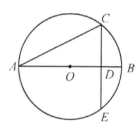

Diagram for question 6

⑦ Given that the length of the base of an isosceles triangle is 6 and the radius of its circumscribed circle is 5, then the length of the two equal sides of the triangle is $\underline{\hspace{2cm}}$.

⑧ Two chords, drawn through a point on the circumference of a circle, are perpendicular to each other. If the distances from the chords to the centre are 2 and 3, then the lengths of the two chords are $\underline{\hspace{2cm}}$.

⑨ In the diagram, AB is a diameter of a circle, centre O, $AB = 4$ and $AC = 2\sqrt{3}$. If D is a point on the circle and $AD = 2$, $\angle DAC = \underline{\hspace{2cm}}$.

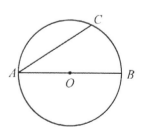

Diagram for question 9

C. Questions that require solutions

10 The diagram shows $\overset{\frown}{AB}$. Use a ruler and a pair of compasses to divide the arc into four equal parts.

Diagram for question 10

11 The diagram shows $\triangle ABC$, with D a point on BC. Draw a circle with point D as its centre and radius CD, intersecting AC and BC at points E and F. $AB = AC = 5$, $\cos B = \dfrac{4}{5}$ and $AE = 1$.

(a) Find the length of segment CD.

(b) Find the distance between point A and point F.

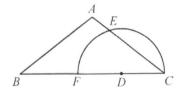

Diagram for question 11

12 The diagram shows an arc-shaped arch bridge in a park. The radius of the circle of which the bridge is part is 10 m and the distance between the top of the bridge, D, and the surface of the water, AB, is $DC = 4$.

(a) Find the width AB, of the surface of the water.

(b) When the water surface rises to EF, the angle of elevation of the top of the bridge, D, from E is α. Given that $\tan \alpha = \dfrac{1}{3}$, find the increase in the height of the water surface.

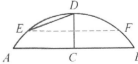

Diagram for question 12

9.8 The relationship between a line and a circle

Learning objective

Understand and apply the relationship between a line and a circle.

A. Multiple choice questions

1 Read these statements.

① A line that shares some common point(s) with a circle is a tangent to the circle.

② A line that is perpendicular to the radius of a circle is a tangent to the circle.

③ If the distance from the centre of a circle to a line is equal to the radius, then the line is a tangent to the circle.

④ A line that passes through an endpoint of the diameter of a circle and is perpendicular to the diameter is a tangent to the circle.

The correct statements are (   ).

A. ① and ②   B. ② and ③

C. ③ and ④   D. ① and ④

2 If the distance from a point on a line to the centre of a circle is equal to the radius of the circle, then the line (   ) the circle.

A. intersects   B. is tangent to

C. intersects or is tangent to   D. none of the above

3 Given that the radius of a circle is 6.5 cm, and the distance between the centre and the line is 5 cm, then there is/are (   ) point(s) of intersection between the line and the circle.

A. 0   B. 1

C. 2   D. uncertain

4 On a coordinate plane, a circle with point (2, 1) as its centre and a radius of 1 must (   ).

A. intersect the x-axis   B. intersect the y-axis

C. be tangent to the x-axis   D. be tangent to the y-axis

B. Fill in the blanks

5 The radius of a circle, centre O, is 10 cm and the distance from its centre O to a line a is d.

① If a is a tangent to the circle, then d = _____.

② If d = 4 cm, then there is/are _____ point(s) of intersection between a and the circle.

③ If d = 14 cm, then there is/are _____ point(s) of intersection between a an the circle.

6 Given that the diameter of a circle, centre O, is 6 cm, point A is on line l and AO = 3 cm, then there is/are _____ point(s) of intersection between line l and the circle.

7 In a right-angled $\triangle ABC$, $\angle C$ = 90°, $\angle A$ = 30° and BC = 6. If AB is a tangent to a circle, centre C, then the radius of the circle is _____.

8 Look at the diagram. $\angle AOB$ = 30°, M is a moving point on OB, and a circle, centre M, has a radius of 2 cm. When OM = _____, OA is a tangent to the circle.

Diagram for question 8

9 In a right-angled $\triangle ABC$, $\angle C$ = 90°, AC = 5 and BC = 12. If a circle with centre C and r as radius is drawn so that there is only one point of intersection between the circle and the hypotenuse AB, then the set of values that radius r can take is _____.

C. Questions that require solutions

10 The diagram shows $\triangle ABC$, with $\angle BAC = 120°$, $AB = AC$ and $BC = 4\sqrt{3}$. If a circle, centre A, has a radius of 2, what is the relationship between BC and the circle?

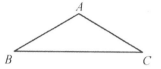

Diagram for question 10

11 The diagram shows a circle, centre O, on a coordinate plane with radius 2. P is a moving point on the circle in the first quadrant. A line passing through P is a tangent to the circle and intersects the x-axis and the y-axis at points A and B, respectively.

(a) Prove that $\triangle OBP$ is similar to $\triangle AOP$.

(b) When P is the midpoint of AB, find the coordinates of point P.

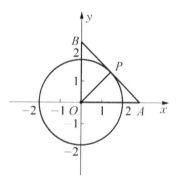

Diagram for question 11

12 In the diagram, the coordinates of point A are $(0, 3)$ and the area of rectangle $ABCO$ is 12. A circle, centre P, moves so that it passes through points A and B. Find the coordinates of the centre, P, when the circle intersects the y-axis and the distance between the two points on the y-axis is 4.

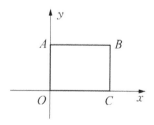

Diagram for question 12

9.9 The relationship between two circles (1)

Learning objective

Understand and apply relationships between two circles.

A. Multiple choice questions

1. Given that the radii of two circles are 3 and 4, and the distance between their centres is 8, then the relationship between the two circles is that (   ).
   A. they touch internally
   B. they intersect at two points
   C. they are apart from each other
   D. they touch externally

2. Given that the radii of two circles with centres O_1 and O_2 are 2 and 1, and the two circles intersect, then the distance O_1O_2 between the centres could be (   ).
   A. 2   B. 4   C. 6   D. 8

3. Given that two circles of radii 3 and 5 have no common point, then the set of values that the distance, d, between the two centres can take is (   ).
   A. $d > 8$
   B. $d > 2$
   C. $0 \leqslant d < 2$
   D. $d > 8$ or $0 \leqslant d < 2$

B. Fill in the blanks

4. Two circles can intersect at _____ points (s), touch each other with _____ common point (s) externally or internally, or be apart externally or internally with _____ common point(s).

5. Given that two circles are tangent to each other and their radii are 5 cm and 4 cm, then the distance between the two centres is _____.

6. In the diagram, a circle, centre P, is tangent to another circle, centre O, internally. Chord AB of the circle with centre O is tangent to the circle, centre P, and $AB \parallel OP$. Given that the radius of the circle, centre O, is 3 and the radius of the circle, centre P, is 1, then the area of quadrilateral $ABPO$ is _____.

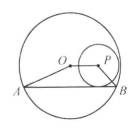

Diagram for question 6

7 The diagram shows two circles, centres A and B, with radii 1 cm and 2 cm, respectively. Both A and B are on line l, and $AB = 6$ cm. Now suppose the circle with centre A is moving to the right along line l at the speed of 1 cm per second, and let the moving time be t seconds. When the two circles intersect, the set of values that t can take is _____.

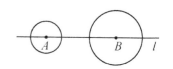

Diagram for question 7

C. Questions that require solutions

8 Given three circles, if every two circles are tangent externally, and the distances between the centres are 6, 8 and 10, respectively, find the radii of the three circles.

9 The diagram shows $\triangle ABC$, in which $\angle C = 90°$, $AC = 12$ and $BC = 8$. Given that AC is a diameter of the circle with centre O, and the circle with centre B has radius 4, prove that the two circles with centres O and B are tangent to each other.

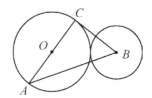

Diagram for question 9

10 The diagram shows a right-angled $\triangle ABC$ and a circle, centre O, that circumscribes $\triangle ABC$. $\angle ACB = 90°$, $AC = 6$ cm and $BC = 8$ cm. Point P is the midpoint of BC. Q is a moving point, starting from point P in the direction of PC at 2 cm/s (Note: Q can be on the extension of PC beyond C.) A circle with point P as its centre has PQ as its radius. Let the time for which point Q is moving be t seconds. Find the value of t when the circle with centre P is tangent to the circle with centre O.

Diagram for question 10

11 The diagram shows a right-angled trapezium $ABCD$, with $AD \parallel BC$, $\angle C = 90°$, $BC = 12$, $AD = 18$ and $AB = 10$. P and Q are two moving points. Starting at the same time, P moves from point D towards A at 2 units per second and Q moves from point B towards C at 1 unit per second. When Q arrives at point C, it stops moving and P stops moving at the same instant. Let the time for which both points move be t (seconds). Given that a circle with BQ as the diameter is tangent externally to the circle with AP as the diameter, then find the value of t.

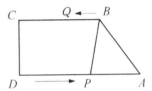

Diagram for question 11

9.10   The relationship between two circles (2)

Learning objective

Understand and apply relationships between two circles.

A. Multiple choice questions

1. If the radius of a circle, centre O_1, is 5, the radius of a circle, centre O_2, is 8 and $O_1O_2 = 4$, then the relationship between the two circles is that (   ).
   A. they have no point of intersection and one is inside the other
   B. they are tangent to each other internally
   C. they intersect at two points
   D. they have no point of intersection and are separate from each other

2. Given that two circles, centres O_1 and O_2, are tangent to each other, the radius of the circle, centre O_1, is 3 cm and the radius of the circle, centre O_2, is 2 cm, then the length of O_1O_2 is (   ).
   A. 1 cm
   B. 5 cm
   C. 1 cm or 5 cm
   D. 0.5 cm or 2.5 cm

3. The three vertices of equilateral $\triangle ABC$ are taken as the centres of three circles. Given that the two circles with centres A and B are tangent to each other externally, the two circles with centres A and C are tangent to each other externally, and the two circles with centres B and C are separate from each other, then statement (   ) about the radii R_A and R_B of the circles with centres A and B, is true.
   A. $R_A > R_B$
   B. $R_A = R_B$
   C. $R_A < R_B$
   D. possibly all of the above

4. Of these statements, (   ) is true.
   A. One circle can be drawn passing through any three points on a plane.
   B. If two circles intersect, then the common chord must be perpendicular to the line connecting the two centres.
   C. Arcs with equal central angles must be equal to each other.
   D. If one circle is tangent to another circle internally, then the distance between the two centres is equal to the sum of the radii of the two circles.

B. Fill in the blanks

5 Two circles, centres O_1 and O_2, are tangent to each other, $O_1O_2 = 8$, and the radius of the circle with centre O_1 is 5. Then the radius of the circle with centre O_2 is _____.

6 The diagram shows a right-angled $\triangle ABC$, with $\angle C = 90°$, $AC = 4$ and $BC = 3$. AB is tangent to a circle, centre C. If another circle, centre A, is tangent to the circle with centre C, the set of values that the radius r of the circle with centre A is _____.

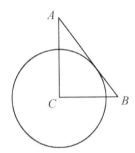

Diagram for question 6

7 The diagram shows a square $ABCD$ with point E on BC. The semicircle, centre E, is tangent externally to the arc with A as its centre and AB as radius. Then $\sin \angle EAB$ = _____.

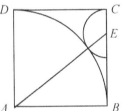

Diagram for question 7

8 The diagram shows right-angled $\triangle ABC$ with $\angle C = 90°$, $AC = 3$ and $BC = 4$, and a circle, centre O, with BC as its diameter. P is a moving point on AC, and the radius of a circle with centre P is 1. Let $AP = x$. If the two circles with centres P and O intersect each other, then the set of values that x can take is _____.

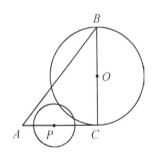

Diagram for question 8

9 Two circles, centres O_1 and O_2, intersect each other at points A and B. If the radii of the two circles are 10 cm and 17 cm, respectively, and the length of the common chord AB is 16 cm, then the distance between the two centres is _____ cm.

C. Questions that require solutions

10 Given that two circles, centres A and B, are tangent to each other, $AB = 12$ and the radius of the circle with centre A is 5, find the radius of the circle with centre B.

11 The diagram shows a rectangle $ABCD$ with side AD along line MN, $BC = 6$ and $AB = 8$. Point E is a moving point on MN. Given that the circle, centre A, with radius AB, is tangent to the circle, centre E, with radius ED, find the radius of the circle with centre E.

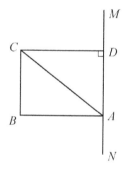

Diagram for question 11

12 The diagram shows point B is on the x-axis of a coordinate plane. A circle, centre B, with radius 3 is tangent to the y-axis, and line l passes through point $A(-2, 0)$ and is tangent to the circle, centre B, intersecting the y-axis at point C. Given that point E is on line l and the circle with centre A and radius AE is tangent to the circle with centre B, find the coordinates of point E.

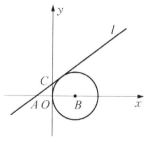

Diagram for question 12

Unit test 9

A. Multiple choice questions

1. If the central angle of a sector is increased to twice its original size and the radius is increased to 3 times its original size, then the area of the sector is increased to (   ) its original size.

   A. 4 times　　　　B. 18 times　　　　C. 12 times　　　　D. 6 times

2. Read these statements about properties of circles. The incorrect one is (   ).

   A. If a diameter of a circle is perpendicular to a chord, then the diameter bisects the chord.

   B. If the distances from the chords to the centre in the same circle are equal, then the minor arcs (or major arcs) of the chords are equal.

   C. If the diameter of a circle bisects a chord, then the diameter is perpendicular to the chord.

   D. If the lengths of two arcs in the same circle or equal circles are equal, then the chords subtended by the arcs are also equal.

3. Inside a circle with radius 10 cm, the set of possible values of d, the distance between any two points, is (   ).

   A. $0 < d < 10$ cm　　B. $0 < d < 15$ cm　　C. $0 \leqslant d < 20$ cm　　D. $10 \leqslant d < 15$ cm

4. The diagram shows a circle with centre O. Chord CD and diameter AB intersect at E, $\angle AEC = 45°$, and $OF \perp CD$ at F. If $OF = 4$ and $DE = 3.5$, then the length of chord CD is (   ).

   A. 7

   C. 15

   B. 7.5

   D. 8.5

   Diagram for question 4

5. The radius of a circle is 6.5 cm and the distance from the centre to a line is 4.5 cm. The line and the circle have (   ) common point(s).

   A. 2　　　　　　B. 0　　　　　　C. 1　　　　　　D. uncertain

213

6 In a circle, the length of an arc is l, the central angle is n, the sector area is S_1, the circumference is C and the area of the circle is S. Then equation (   ) is not true.

A. $\dfrac{l}{C} = \dfrac{n}{360}$ B. $\dfrac{l}{S} = \dfrac{C}{S_1}$ C. $\dfrac{l}{C} = \dfrac{S_1}{S}$ D. $\dfrac{S_1}{S} = \dfrac{n}{360}$

7 The diagram shows a right-angled trapezium $ABCD$ with $AD \parallel BC$ and $\angle A = 90°$. Line segment AB is a tangent to a circle with diameter CD at point E. If $AD = 3$ and $BC = 4$, then the radius of the circle, centre O, is (   ).

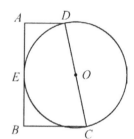

A. 7

B. 3.5

C. $5\sqrt{2}$

D. none of these

Diagram for question 7

B. Fill in the blanks

8 Given that the central angle of a sector is $45°$ and the radius is 8 cm, then its perimeter is _____ cm.

9 Given that the radius of an arc is 4 cm and its central angle is $72°$, then the arc length is _____ cm.

10 The length of the hour hand of a clock is 9 cm. The area that the hour hand sweeps out on the clock face from 6 o'clock to 10 o'clock in the morning is _____ cm^2.

11 In the diagram, AB and CD are two diameters of a circle with centre O, $\overparen{AE} = \overparen{AC}$ and $\angle AOE = 32°$. Then $\angle COE$ is _____ degrees.

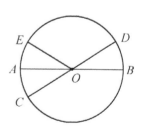

Diagram for question 11

12 The longest chord in a circle is _____.

13 In a circle, the angle subtended by an arc at the circumference is _____ the angle subtended at the centre.

14 The diagram shows concentric circles. Chord AB in the big circle is trisected (divided into three equal parts) by the small circle. OP is the distance between the centre and chord. Given that $PD = 2$ cm, then the length of BC is _____ cm.

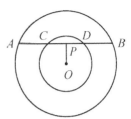

Diagram for question 14

15 Consider the four pairs of measurements in a given circle or two equal circles: two central angles, two major arcs (or minor arcs), two chords, and the distances from the chords to centre(s). If the two in one pair are equal, then the two in the other three corresponding pairs are _____.

16 The diagram shows a circle with centre O. CD is a diameter, $\angle ACD = 25°$ and $OA \perp OB$. Then $\angle AOD =$ _____ and $\angle BOD =$ _____.

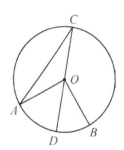

Diagram for question 16

17 The diagram shows a circle, centre O. Chord CD intersects diameter AB at point E and $\angle AEC = 30°$. Given that $OF \perp CD$ at F, $OF = 2$ and $DE = 3$, then $DC =$ _____.

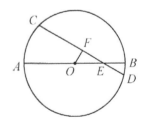

Diagram for question 17

18 In $\triangle ABC$ in the diagram, $AB = 4$, $AC = 5$ and $BC = 6$. Three circles, centres A, B and C, are tangent to each other externally. The radius of the circle with centre A is _____.

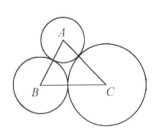

Diagram for question 18

C. Questions that require solutions

19 The diagram shows a circle with centre O. The midpoints of the two chords AB and CD are M and N, respectively, and $\angle OMN = \angle ONM$. Prove that $AB = CD$.

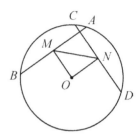

Diagram for question 19

20 The diagram shows a semicircle with centre O and diameter AB. $\overset{\frown}{CD} = \overset{\frown}{DE} = \overset{\frown}{EF} = \overset{\frown}{FB}$, and $\angle AOC = 60°$.

(a) Find the size of $\angle FOB$.

(b) Prove that $OE \perp CB$.

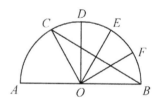

Diagram for question 20

21 The diagram shows a circle with centre O and diameter EF. $\angle AOB = 90°$, $EF \perp AB$ at C. and $OC = 40$ cm.

(a) Find the length of chord AB.

(b) Find the size of $\angle AEB$.

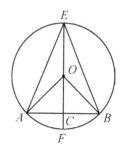

Diagram for question 21

22 The diagram shows a square $ABCD$, with $ED = DA = AF = 2$ cm. What is the area of the shaded region?

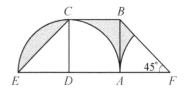

Diagram for question 22

23 The diagram shows a rectangle $ABCD$ with $AB = 3$ and $BC = 4$. P is a point on the extension BC beyond C. Line segment AP intersects CD at point E. Q is the image of E after reflection of AP in line AD. Let $CP = x$, and $DQ = y$.

(a) Prove that $\triangle ADQ \backsim \triangle PBA$, and find an algebraic expression of y as a function of x.

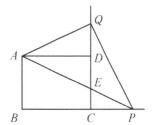

Diagram for question 23

(b) When point P is moving, will the area S of $\triangle APQ$ be affected? If it won't, find the value of S. If it will, give your reasons.

(c) When a circle, centre Q, with radius 4 is tangent to line AP, and another circle, centre A is tangent to the circle with centre Q, find the radius of the circle, centre A.

End of year test

A. Multiple choice questions (24%)

1 Given that $ax - 6 = 3x + 2y$ is a linear equation in two variable x and y, then (   ).

A. $a \neq 3$ 　　　　　　　　　　　B. $a \neq -3$

C. $a \neq 0$ 　　　　　　　　　　　D. $a \neq 6$

2 Of the sets of simultaneous equations:

① $\begin{cases} x - 2 = 0, \\ x + 3y = 6 \end{cases}$ 　　② $\begin{cases} x + y = 4, \\ x - z = 1 \end{cases}$ 　　③ $\begin{cases} x - y = 2, \\ x + y = 4, \\ 2x - 3y = 1 \end{cases}$ 　　④ $\begin{cases} x - 3 = y, \\ xy + 1 = 4 \end{cases}$,

(   ) are simultaneous linear equations in two variables.

A. one 　　　　　B. two 　　　　　C. three 　　　　　D. four

3 The correct representation of the solution set to the inequality $2(x + 1) < 3x$ on the number line is (   ).

A.
$$-2\ -1\ 0\ 1\ 2\ 3$$

B.
$$-2\ -1\ 0\ 1\ 2\ 3$$

C.
$$0\ 1\ 2\ 3\ 4\ 5$$

D.
$$0\ 1\ 2\ 3\ 4\ 5$$

4 The simplest common denominator of the algebraic fractions $\dfrac{3}{4y}$, $\dfrac{2}{3xy^2}$ and $\dfrac{1}{6x^2}$ is (   ).

A. $12xy^2$ 　　　　　　　　　　B. $12x^2y^2$

C. $24x^2y^2$ 　　　　　　　　　　D. $24x^3y^3$

5 In trapezium $ABCD$, $AD \mathbin{/\!/} BC$ and $AB = CD$. The correct conclusion is (   ).

A. $\overrightarrow{AB}$ and $\overrightarrow{DC}$ are equal vectors.

B. $\overrightarrow{AC}$ and $\overrightarrow{BD}$ are equal vectors.

C. $\overrightarrow{AD}$ and $\overrightarrow{CB}$ are inverse vectors.

D. $\overrightarrow{AD}$ and $\overrightarrow{CB}$ are parallel vectors.

6 The diagram shows a vector on a coordinate plane.

The incorrect representation of the vector is (   ).

A. $\overrightarrow{OP}$

B. $\begin{pmatrix} 5 \\ -3 \end{pmatrix}$

C. **a**

D. $\begin{pmatrix} -3 \\ 5 \end{pmatrix}$

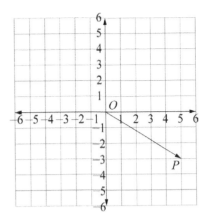

Diagram for question 6

7 Inside a rectangle 5 cm long and 2 cm wide, there is an arc with the centre at one of the vertices and a radius of 2 cm. The arc length is (   ). (Take π as 3.14)

A. 4.71 cm   B. 3.14 cm   C. 6.28 m   D. 12.56 cm

8 The radii of two circles, centres O_1 and O_2, are 3 cm and 4 cm respectively. Given that $O_1O_2 = 1$ cm, then the two circles (   ).

A. have no common point   B. are tangent to each other externally

C. intersect at two points   D. are tangent to each other internally

B. Fill in the blanks (22%)

9 Given that the radius of a sector is 3 cm and the central angle is $60°$, then the area of the sector is _____ cm^2. (Give your answer in terms of π.)

10 There are _____ solution sets to the linear equation in two variables: $x + 2y = 7$. When x and y are positive integers, the solutions are _____.

11 Given that $\begin{cases} x = -1 \\ y = 2 \end{cases}$ and $\begin{cases} x = 2 \\ y = 3 \end{cases}$ are both the solutions to the equation $ax + by = 21$, then

$a =$ _____ and $b =$ _____.

12 Given that $a^2 - b^2 = 15$ and $a + b = 5$, then the value of $a - b$ is _____.

13 Simplifying $(x^{-3}yz^{-2})^2$, the result expressed in exponent form with only positive integers is _____.

14 Calculate: $\dfrac{a + 2b}{6ab} - \dfrac{a - b}{6ab} = $ _____.

15 The solution to the equation $x^2 = 3x$ is _____.

16 The smallest integer solution to the inequality $\dfrac{1 - 2x}{3} - 1 < 0$ is _____.

17 Calculate: $\overrightarrow{AB} - \overrightarrow{CB} + \overrightarrow{CA} = $ _____.

18 When a column vector $\mathbf{a} = \begin{pmatrix} -1 \\ 1 \end{pmatrix}$ is translated 3 units to the right and 3 units downwards, the resultant vector is _____ and the magnitude is _____.

19 The diagram shows a circle, centre O. Point C is on the circle and $\overset{\frown}{AB} = \overset{\frown}{BD}$. If $\angle ACB = 35°$, then $\angle AOB = $ _____ and $\angle ADB = $ _____.

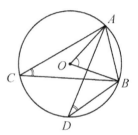

Diagram for question 19

C. Short answer questions (24%)

20 Factorise: $(2a - b)^3 - (a + 4b)^2(2a - b)$.

21 Solve the equation: $\dfrac{x^2}{x + 1} - x - 1 = 0$.

22 Simplify first and then evaluate: $\dfrac{x}{x + 1} - \dfrac{x + 3}{x + 1} \times \dfrac{x^2 - 1}{x^2 + 2x - 3}$ for $x = \dfrac{4}{3}$.

23 Calculate: $(2)(7^{\frac{3}{2}} \times 49^{-\frac{3}{4}})^{\frac{1}{3}}$.

24 Calculate: $\left(3\sqrt{18} + \frac{1}{5}\sqrt{50} - 4\sqrt{\frac{1}{2}}\right) \div \sqrt{32}$.

25 In the diagram, the circle with centre E passes through three points A, C and D and the circle with centre D passes through three points B, F and E. Given that $\angle A = 63°$, find the value of θ.

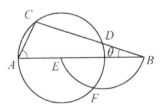

Diagram for question 25

26 Solve the equation $3x^2 - 8x + 3 = 0$ by completing the square.

27 Solve the simulatenous equations: $\begin{cases} x^2 + y^2 = 125 \\ 2x - y = 0 \end{cases}$.

D. Questions that require solutions (30%)

28 $x^2 - mx - 2 = 0$ is a quadratic equation in x.

(a) Find the value of m and the other root of the equation when $x = -1$ is one root of the equation.

(b) Given that m is a real number, determine the nature of the roots of the equation. Give a reason for your answer.

29 The diagram shows $\triangle ABC$ with point D the midpoint of AC. Let $\overrightarrow{AD} = \mathbf{a}$ and $\overrightarrow{BD} = \mathbf{b}$.

(a) Use vectors $\mathbf{a}$ and $\mathbf{b}$ to express vectors: $\overrightarrow{AB} = $ _____ , $\overrightarrow{CB} = $ _____ .

(b) Draw: $\overrightarrow{BD} + \overrightarrow{AC}$ and $\overrightarrow{BD} - \overrightarrow{AC}$.

(Just draw them, there is no need to write the steps.)

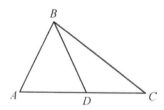

Diagram for question 29

30 In the diagram, $\triangle ABD$ and $\triangle BCD$ are equilateral triangles, both with side length 3 cm. Arc BD is centred at A with AB as the radius. Arc CD is centred at B with BC as the radius. Find the perimeter of the shaded region. (Take π as 3. 14)

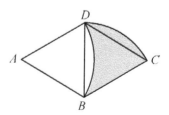

Diagram for question 30

31 A car drives from town A to town B and is expected to arrive at a given time. If the car travels at 40 km/h, it will arrive at the destination 30 minutes later than the given time. If it travels at 50 km/h, it will arrive 30 minutes earlier than the given time. Find the distance between towns A and B and the expected time taken for the journey.

32 The cost price of a table is £100 per unit. When the shop sells these tables at £120 each, it can sell 500 of them. Each time the selling price is raised by £1, the shop sells 10 fewer tables. If the shop plans to earn a total profit of £12 000, then at what price should the shop sell the tables? How many tables should the shop purchase at the cost price?

33 In Figure 1, the radius of the circle, centre O, is 3. A is a fixed point on the circle and P is a moving point on the circle, different from point A.

(a) Find the length of AP when $\tan A = \dfrac{1}{2}$.

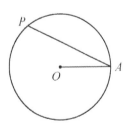

Figure 1

(b) Another circle, centre Q, passes through points P and O, and point Q is on the line segment AP (as shown in Figure 2). Let $AP = x$ and $QP = y$. Find an expression of y as a function of x, and write the domain of the function (the set of values that x can take).

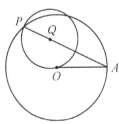

Figure 2

Diagram for question 33

Answers

Chapter 1   Introduction to linear inequalities

1.1   Inequalities and their properties (1)

1 D **2** B **3** the same number, unchanged

4 $a + (-b) > 0$ **5** $8 - 2y \geqslant 0$ **6** $\dfrac{x}{2} - 3 \leqslant -5$ **7** $a^2 + b^2 \geqslant 4ab$ **8** (a) $<$ $>$ (b) $<$ (c) $\leqslant$ (d) $>$ (e) $<$ **9** $a - b < b - a$ **10** $\dfrac{a}{9} - 8 > \dfrac{a}{9} - 9$ **11** $\dfrac{x^2 - y^2 + 1}{2} > \dfrac{x^2 - 2y^2 + 1}{3}$ **12** If $a > -2$, then $\dfrac{2}{3}a + 5 > 3 - \dfrac{1}{3}a$; if $a = -2$, then $\dfrac{2}{3}a + 5 = 3 - \dfrac{1}{3}a$; if $a < -2$, then $\dfrac{2}{3}a + 5 < 3 - \dfrac{1}{3}a$

1.2   Inequalities and their properties (2)

1 D **2** A **3** D **4** the same positive number **5** the same negative number **6** (a) $>$ (b) $<$ (c) $<$ (d) $>$ **7** (a) $>$ (b) $>$ (c) $<$ (d) $>$ (e) $<$ **8** (a) $a - c < 2b - c$ (b) $ac^2 \leqslant 2bc^2$ (c) If $c > 1$, then $\dfrac{a}{c - 1} < \dfrac{2b}{c - 1}$; if $c < 1$, then $\dfrac{a}{c - 1} > \dfrac{2b}{c - 1}$ **9** $a - b < -\sqrt{a^2} < -b < -a$ **10** $m < mn^2 < mn$

1.3   Solve linear inequalities in one variable (1)

1 B **2** B **3** D **4** B **5** the values that the variable can take to make the inequality true **6** solution set **7** solution set to the inequalities **8** one, infinitely many **9** $a < 3$ **10** (a) $x \geqslant -5$ (b) $x < 1$ (c) $x < -5$ (d) $x > -\dfrac{2}{3}$ (e) $x \geqslant \dfrac{2}{3}$ (f) $x \leqslant -\dfrac{2}{3}$ **11** (a) $x < \dfrac{9}{10}$ (b) $x \geqslant 3$ (c) $x > -5.6$

(d) $x \leqslant -3$ **12** $a = 1, 6$ **13** (a) $40 + (31 - 3 - 10)x > 176$ (b) The minimum value is 8.

1.4   Solve linear inequalities in one variable (2)

1 C **2** B **3** C **4** variable, degree **5** 3 **6** $x = 0, 1, 2, 3$ **7** 1, 2, 3 or 2, 3, 4 or 3, 4, 5 **8** $6 \leqslant a < 8$ **9** (a) $x < -2$ (b) $x \leqslant -\dfrac{5}{2}$ (c) $x > -2$ **10** The range of value a is: $a < 2$, and the result of simplification is: a **11** 12 **12** $a = 3$

1.5   Solve linear inequalities in one variable (3)

1 A **2** B **3** $x > \dfrac{4}{7}$ **4** $x \leqslant \dfrac{3}{2}$ **5** $y > 2$ **6** $m \leqslant -\dfrac{9}{17}$ **7** $m > 3$ **8** (a) $x > \dfrac{16}{11}$ (b) $x < 3$ **9** $x = -1$ **10** $m \leqslant -1$ **11** $k > 1$ **12** $a = -2$

1.6   Simultaneous linear inequalities in one variable (1)

1 C **2** D **3** D **4** linear inequalities in one variable **5** $9 \leqslant x \leqslant 12$ **6** $x > 6$ **7** $-2 \leqslant x < 3$ **8** $a < 3$ **9** (a) There is no solution. (b) $-1 \leqslant x < 1$ **10** $x = 0, 1, 2, 3$ **11** $a = -2, b = \dfrac{8}{3}$ **12** $a \leqslant -4$

1.7   Simultaneous linear inequalities in one variable (2)

1 C **2** D **3** A **4** $b < x < a$ **5** $c < x < b$ **6** -3 **7** $x = 1$ **8** -9 **9** $x = -1, 0, 1, 2$ **10** (a) $-6 \leqslant x \leqslant 5$ (b) There is no solution **11** 28 children **12** The perimeter is: 39 or 44

Unit test 1

1 C **2** C **3** B **4** B **5** A **6** $>$

⑦ $a > \dfrac{1}{3}$　⑧ $x > 2$　⑨ $> -\dfrac{1}{3}$　⑩ $9 \leqslant$

$m < 12$　⑪ $-\dfrac{1}{2} < x \leqslant \dfrac{7}{2}$　⑫ -7　⑬ $x <$

10　⑭ $x < 7$　⑮ $x = -2, -1, 0$

⑯ (a) The second car park; £0.50　(b) more

than 2 hours　⑰ 12　⑱ $9 < a \leqslant 12$

Chapter 2　Simultaneous linear equations

2.1　Linear equations in two variables

① B　② C　③ B　④ an equation that has

two variables and the highest degree of any term is

one　⑤ both, equal　⑥ 6　⑦ $-\dfrac{1}{3}$

⑧ $\begin{cases} x = -6, \\ y = 6 \end{cases}$　⑨ (a) $y = \dfrac{3x - 9}{5}$　(b) $x =$

$\dfrac{5y + 9}{3}$　⑩ $\begin{cases} x = 1, \\ y = 15, \end{cases} \begin{cases} x = 2, \\ y = 11, \end{cases} \begin{cases} x = 3, \\ y = 7, \end{cases}$

$\begin{cases} x = 4, \\ y = 3 \end{cases}$　⑪ $\begin{cases} x = -7, \\ y = -1, \end{cases} \begin{cases} x = -3, \\ y = -4 \end{cases}$　⑫ $y =$

$5x + 9$

2.2　Simultaneous linear equations in two variables and their solutions (1)

① C　② two, one　③ the solution that

satisfies all the equations in the system

④ substitute, one variable　⑤ $-1, 1$　⑥ 1

⑦ (a) $\begin{cases} x = 3, \\ y = -2 \end{cases}$　(b) $\begin{cases} x = 2, \\ y = 1 \end{cases}$

⑧ (a) $\begin{cases} x = \dfrac{1}{2}, \\ y = -\dfrac{1}{2} \end{cases}$　(b) $\begin{cases} x = -2, \\ y = 1 \end{cases}$

⑨ (a) $\begin{cases} x = 2, \\ y = -1\dfrac{10}{11} \end{cases}$　(b) $\begin{cases} x = 2.123, \\ y = -4.45 \end{cases}$

⑩ $m = -2$　⑪ $\begin{cases} a = 1, \\ b = 2 \end{cases}$

2.3　Simultaneous linear equations in two variables and their solutions (2)

① D　② D　③ substitution method,

elimination method　④ -14　⑤ $-2, -5$

⑥ -19　⑦ 4　⑧ (a) $\begin{cases} x = \dfrac{1}{3}, \\ y = -2 \end{cases}$

(b) $\begin{cases} x = 11, \\ y = -7 \end{cases}$　⑨ (a) $\begin{cases} x = 5, \\ y = 1 \end{cases}$　(b) $\begin{cases} x = 7, \\ y = 8 \end{cases}$

⑩ (a) $\begin{cases} x = 11, \\ y = 11 \end{cases}$　(b) $\begin{cases} x = 3, \\ y = -1 \end{cases}$　⑪ $k = 2,$

$\begin{cases} x = 3, \\ y = -3 \end{cases}$

2.4　The application of simultaneous linear equations (1)

① Class A has planted 36 trees and Class B has

planted 42 trees.　② 2580 kg, 3150 kg　③ 8

workers are to make type-A components and 18

workers are to make type-B components

④ Group A has 32 students, Group B has 36

students　⑤ 32 students　⑥ The speed of the

train: 20 m/s; the length of the train: 200 m

⑦ 1200 m 404 trees　⑧ In the outward trip,

the length of the uphill road: 42 km, the flat

roads: 30 km, the downhill roads: 70 km

2.5　The application of simultaneous linear equations (2)

① 9 people　② Year 6: 160 students, Year 7:

156 students, Year 8: 104 students　③ Alex

made 50 components each day, and Bobbie made

60 each day.　④ The two-digit number is 21

⑤ 150 square tables　⑥ Person A can make 20

spare parts each day and Person B can make 30

each day.　⑦ Type A: 30, type B: 60 [Hint:

Let type A and type B be x and y, therefore the

equations are $\begin{cases} x + 2y = 150, \\ 4x + 3y = 300 \end{cases}$ and the solutions

are: $\begin{cases} x = 30, \\ y = 60 \end{cases}$]

Unit test 2

① D　② C　③ C　④ D　⑤ C

⑥ $\dfrac{25}{3}$　⑦ $y = \dfrac{3x - 12}{2}$　⑧ $\begin{cases} x = 0, \\ y = 5 \end{cases} \begin{cases} x = 2, \\ y = 2 \end{cases}$

⑨ $\dfrac{85}{36}$　⑩ 3　⑪ $-2, 3$　⑫ 6　⑬ $\dfrac{1}{4}$

⑭ $m = 1, 2, 3$　⑮ $\begin{cases} x = 1, \\ y = -1 \end{cases}$　⑯ $\begin{cases} x = -4, \\ y = 7 \end{cases}$

⑰ $x < -2$　⑱ $40, 80$　⑲ A travelled $\frac{27}{4}$ km/h, and B travelled $\frac{9}{4}$ km/h　⑳ Winning matches: 5, drawn matches: 4, lost matches: 2

Chapter 3　Laws of indices and factorisation

3.1　Power of a power

① C　② C　③ C　④ B　⑤ a^{36}
⑥ a^{12}　⑦ $2a^6$　⑧ a^{30}　⑨ $-(a-b)^{11}$
⑩ 2^{4n}　⑪ a^{120}　⑫ $-(x-y)^{8m+5}$
⑬ $3a^{2n}b^{6n}$　⑭ $\frac{5}{2}$　⑮ a^2; $a^2 + b^3 + a^2b^3$

3.2　Power of a product

① C　② B　③ A　④ C　⑤ $2ab^3$
⑥ $16; 40$　⑦ $-\frac{27}{64}a^3b^6$　⑧ $36x^6$　⑨ $-\frac{10}{3}$
⑩ $2^{mp}a^{np}b^{2p}$　⑪ $4.$　⑫ $-12x^5y^3$
⑬ $-a^{3n}b^{9n} - 23a^{6n}b^{9n}$　⑭ 64　⑮ 100^{99} (or 10^{198})

3.3　Factorisation using formulae (1): difference of two squares

① C　② B　③ D　④ $(x+3)(x-3)$
⑤ $\left(2x + \frac{1}{3}y\right)\left(2x - \frac{1}{3}y\right)$　⑥ $\left(\frac{1}{3}b + a\right)$ $\left(\frac{1}{3}b - a\right)$　⑦ $3x(x+3)(x-3)$　⑧ $(a^2 + 2)(a^2 - 2)$　⑨ $(4x + 7y^2)(4x - 7y^2)$
⑩ $2y(x-y)$　⑪ $2x(1+2x)(1-2x)$
⑫ $(a+b+10)(a+b-10)$　⑬ $7mn(m + 2n)(m - 2n)$　⑭ $3m(x-y)(x-y+3m)(x - y - 3m)$

3.4　Factorisation using formulae (2): completing the square

① C　② C　③ C　④ $\left(x + \frac{1}{2}\right)^2$
⑤ $\left(\frac{1}{2}x - 1\right)^2$　⑥ $(2x-1)^2$　⑦ $(xy - 2)^2$
⑧ $(3-2t)^2$　⑨ $2x; 2x + 5y$　⑩ $4m^2n^2$; $2mn + 1$　⑪ $(3x + 7y)^2$　⑫ $(a - b - c)^2$

⑬ $-\left(a^2b - \frac{1}{2}x\right)^2$　⑭ $(2 - 3x + 3y)^2$

3.5　Factorisation using formulae (3): revision and practice

① B　② B　③ C　④ $\frac{1}{4}n^2$, $m + \frac{1}{2}n$
⑤ $\left(4a + \frac{1}{3}b\right)\left(4a - \frac{1}{3}b\right)$　⑥ $\left(\frac{1}{2}y + 1\right)^2$
⑦ $(x + y + 3)^2$　⑧ $(x+1)^4$　⑨ $x(x-3)^2$
⑩ $(x + 1 + y)(x + 1 - y)$　⑪ $(ab - 4c)^2$
⑫ $(2x + 3y)^2(2x - 3y)^2$　⑬ $\frac{1}{32}$　⑭ (a) $n^2 - (n+2)^2 = -4(n+1)$　(b) $100 - 144 = -44$

3.6　Factorisation using the cross-multiplication method

① A　② D　③ B　④ $9x$; $x + 5$　⑤ x; $x - 5$　⑥ 2　⑦ $-6; 4$　⑧ $(x-3)(x-2)$
⑨ $(x+4)(x-3)$　⑩ $(x-3)(x+2)$
⑪ $(x+2)(x-2)(x+1)(x-1)$　⑫ $(x + y - 6)(x + y + 2)$　⑬ $(xy - 8)(xy - 2)$
⑭ $a = -1, 1, -5$ or 5

3.7　Factorisation by grouping

① B　② A　③ $ab + b^2$; $a + b, b - c$
④ $ax - b$; $ax - b$; $x + 1$　⑤ $2ax - ay$; $4bx - 2by$; $2x - y$; $a + 2b$　⑥ x^2; $a^2 + 2ab + b^2$; $x + a + b$; $x - a - b$　⑦ $(a+b)(a - b + x)$
⑧ $(y-1)(x-1)$　⑨ $(x + 2y - 1)(x - 2y + 1)$
⑩ $(x+a)(x-a-2)$　⑪ $(x-3)(7x+y)$
⑫ $(2a - b + 4)(2a - b - 4)$
⑬ $(x - 2y - 3)(x - 2y + 1)$
⑭ $\left(m - \frac{1}{2} + n\right)\left(m - \frac{1}{2} - n\right)$

Unit test 3

① D　② D　③ C　④ C　⑤ B　⑥ A
⑦ m^8　⑧ x^{12}　⑨ $-\frac{8}{27}x^3y^9$　⑩ $7a^4x^8$
⑪ $2xy^2$　⑫ $a(x+5)(x-5)$　⑬ $x(y-10)^2$
⑭ $x(x-3)(x+1)$　⑮ $(a + x - 1)(a - x - 1)$
⑯ 6　⑰ $8(a-b)^2(a+b)$　⑱ $(x+3)(a-2)$
⑲ -10^{2017}　⑳ 265　㉑ $2a(x+1)(x-1)$ $(x+2)(x-2)$; $x = 1, -1, 2,$ or -2

Chapter 4　Algebraic fractions

4.1　The meaning of algebraic fractions

① C　**②** A　**③** C　**④** C　**⑤** For

example: $\dfrac{a}{2}$; $\dfrac{1}{a}$ (to write one each)　**⑥** (a) $a \div$

$2a - b$　(b) $-(a - b) \div (2x - y)$

⑦ (a) $\dfrac{a + 2}{3(b - 5)}$　(b) $-\dfrac{x + 1}{x}$　**⑧** $\dfrac{ab}{a + b}$

⑨ $\dfrac{5}{4}$　**⑩** 2 or –2　**⑪** $\dfrac{3}{4}$　**⑫** (a) 5

(b) Does not exist　**⑬** $x = \dfrac{3}{2}$　**⑭** $-\dfrac{3}{2} <$

$x < 2$ $\left[\text{Hint}: \begin{cases} x - 2 > 0, \\ 3 + 2x < 0 \end{cases} \text{or} \begin{cases} x - 2 < 0, \\ 3 + 2x > 0 \end{cases}\right]$

⑮ $m = -8, 2, 4$ or 10　$\left[\text{Hint}: \dfrac{2m + 7}{m - 1} = 2 + \right.$

$\left. \dfrac{9}{m - 1} \right]$

4.2　Properties of algebraic fractions

① D　**②** C　**③** D　**④** A　**⑤** (a) $2x^2 y$

(b) $2x$　(c) $2x^2$　(d) $5xy - 5y^2$　(e) $3x - y$;

$3x^2 + xy$　**⑥** 3　**⑦** 3 or 4　**⑧** 0　**⑨** 0

⑩ (a) $-3yz^{10}$　(b) $\dfrac{a + 7}{a - 3}$　(c) $\dfrac{a}{a^n + b^n}$

(d) $\dfrac{x - y}{x + y}$　(e) $\dfrac{a + b - c}{a - b - c}$　**⑪** 2 or –1

4.3　Multiplying and dividing algebraic fractions (1)

① C　**②** A　**③** D　**④** D　**⑤** new

numerator; new denominator; upside down;

multiply　**⑥** $-\dfrac{2}{3ac}$　**⑦** $\dfrac{15ab}{a + b}$　**⑧** $-\dfrac{a^2 + a}{a - 1}$

⑨ $\dfrac{x^5}{x - 2}$　**⑩** 6　**⑪** $\dfrac{x^2 - 2xy}{x + y}$　**⑫** $-\dfrac{1}{2x - 4}$

⑬ $a^2 - a - 2$; –3　**⑭** 24

4.4　Multiplying and dividing algebraic fractions (2)

① D　**②** A　**③** D　**④** C　**⑤** $\dfrac{9\,a^2 x^6}{16\,p^2 m^6}$

⑥ $-3x^3 y$　**⑦** $-\dfrac{a}{b}$　**⑧** $-\dfrac{25\,y^7}{4\,x^{15}}$

⑨ $\dfrac{(x - y)^2}{x + y}$　**⑩** $\dfrac{1}{xy}$　**⑪** $\dfrac{x + 2}{x + 1}$

⑫ $-\dfrac{x^2}{x - 2}$　**⑬** $\dfrac{n^3(n - m)}{m}$; 2

4.5　Adding and subtracting algebraic fractions (1)

① B　**②** A　**③** B　**④** B　**⑤** $-\dfrac{1}{a}$

⑥ $-\dfrac{1}{x + 3}$　**⑦** $\dfrac{6}{x - 2}$　**⑧** $\dfrac{1}{(x - y)^2}$

⑨ $\dfrac{x + y}{x - y}$　**⑩** $-a - 3b$　**⑪** $\dfrac{x + 1}{x - 2}$　**⑫** 0

⑬ 0

4.6　Adding and subtracting algebraic fractions (2)

① D　**②** B　**③** C　**④** $\dfrac{a + b}{ab}$　**⑤** $\dfrac{1 - 2ab}{2a^2}$

⑥ $\dfrac{2b}{4a^2 - b^2}$　**⑦** $\dfrac{y^2}{x + y}$　**⑧** $\dfrac{a^2 + 1}{a^2 - a}$　**⑨** 1; –1

⑩ $\dfrac{x - 5}{6 - 2x}$　**⑪** $\dfrac{1}{2x - 4}$　**⑫** 1

⑬ $-\dfrac{x(x - 2)}{x - 1}$; $\dfrac{8}{3}$

⑭ $\dfrac{-10x + 10}{(x + 1)(x + 2)(x - 3)(x - 4)}$ $\left[\text{Hint}: \text{the}\right.$

original expression $= \left(1 + \dfrac{1}{x + 1}\right) - \left(1 + \dfrac{1}{x + 2}\right) -$

$\left(1 - \dfrac{1}{x - 3}\right) + \left(1 - \dfrac{1}{x - 4}\right) = \left(\dfrac{1}{x + 1} + \dfrac{1}{x - 3}\right) -$

$\left(\dfrac{1}{x + 2} + \dfrac{1}{x - 4}\right) = (2x - 2)\left(\dfrac{1}{(x + 1)(x - 3)} - \right.$

$\left. \left. \dfrac{1}{(x + 2)(x - 4)}\right)\right]$

4.7　Adding and subtracting algebraic fractions (3)

① D　**②** D　**③** A　**④** $\dfrac{3ac^2 + 2b^2}{6a^2 bc}$

⑤ $\dfrac{2x^2 + 3}{x - 1}$　**⑥** $\dfrac{1}{a + 1}$　**⑦** $\dfrac{x - y}{xy}$　**⑧** 1; 2

⑨ $-\dfrac{x + 2}{x + 1}$　**⑩** $x - 1$　**⑪** $\dfrac{2}{3}$　**⑫** $\dfrac{1}{a^2 + 2a}$; 1

⑬ $A = -2, B = 4, C = 4$ $\left[\text{Hint}: \dfrac{6x^2 + 2x + 4}{x(x - 1)(x + 2)} = \right.$

$\dfrac{(A + B + C)x^2 + (A + 2B - C)x - 2A}{x(x - 1)(x + 2)}$, then

$\begin{cases} A + B + C = 6, \\ A + 2B - C = 2, \\ -2A = 4, \end{cases}$ the solutions are $\begin{cases} A = -2, \\ B = 4, \\ C = 4 \end{cases}$]

4.8 Algebraic fraction equations that can be transformed to linear equations in one variable

❶ B ❷ C ❸ A ❹ 5 ❺ $\dfrac{7}{3}$ ❻ $x =$ -2 ❼ 8 ❽ $-\dfrac{25}{6}$ ❾ 6 ❿ no solution

⓫ $x = 12$ ⓬ $x = \dfrac{11}{4}$ ⓭ $x = -\dfrac{9}{2}$ [Hint:

$\dfrac{1}{x + 2} - \dfrac{1}{x + 3} = \dfrac{1}{x + 6} - \dfrac{1}{x + 7}$]

4.9 Integer exponents and their operations (1)

❶ D ❷ D ❸ D ❹ (a) $3x^{-2}y^{-3}$

(b) $2x^{-1} + 3y^{-2}$ (c) $4^{-1}(x - y)^3(x + y)^{-5}$

(d) $(2x - y)x^{-5}y^{-1}$ ❺ (a) $\dfrac{5}{x^2}$ (b) $\dfrac{a^3b^5x}{45y^4}$

(c) $\dfrac{xy}{x + y}$ (d) $-\dfrac{3xy^2}{(x + y)^2}$ ❻ $1\left(x \neq \dfrac{1}{2}\right)$

❼ $\dfrac{17}{4}$ ❽ 2 ❾ $\dfrac{y - x}{x + y}$ ❿ $\dfrac{63}{4}$ ⓫ -6

⓬ 0 ⓭ 3

4.10 Integer exponents and their operations (2)

❶ D ❷ D ❸ C ❹ (a) -5.768×10^{-1} (b) -1.00109×10^2 (c) 4.3×10^{-4}

(d) 3.67×10^{-3} (e) 3.6799×10^6 (f) 1.24×10^{-3} (g) 6.3×10^{-5} ❺ (a) 5 030 000

(b) -0.000315 (c) 0.000 010 9

(d) -0.00423 ❻ -8 ❼ 23 ❽ -4.8×10^{-9} ❾ $\dfrac{b - a}{ab}$ ❿ $\dfrac{3a^2 + 5ab - b^2}{4a^2 - 4ab + b^2}$ ⓫ 0

[Hint: $x^2 + \dfrac{1}{x^2} = 2$, $x^4 + \dfrac{1}{x^4} = 2$, $\cdots$, $x^{2048} +$

$\dfrac{1}{x^{2048}} = 2$, $x^{2048} + \dfrac{1}{x^{2048}} - 2 = 0$] ⓬ $\dfrac{25}{27}$

Unit test 4

❶ B ❷ C ❸ B ❹ A ❺ D ❻ D

❼ $a^2 - 1$ or $1 - a^2$ ❽ $\neq -2$ ❾ $= 1$

❿ 4.7×10^{-4} ⓫ $-\dfrac{a}{2b}$ ⓬ $\dfrac{1}{x - 3}$ ⓭ 11

⓮ $\dfrac{3}{2}$ ⓯ 0 ⓰ $-6a^4b^{-5}$ ⓱ $-\dfrac{9}{2}$

⓲ no solution ⓳ $\dfrac{1}{10}$ ⓴ $\dfrac{3b}{64a^{11}}$ ㉑ 0

㉒ 1 ㉓ $\dfrac{5}{3}$ ㉔ $x = \dfrac{5}{2}$ ㉕ $\dfrac{x}{x + 1}$; -1

㉖ $A = -1$, $B = 1$ ㉗ 6 km/hour

㉘ $\dfrac{3x - 4}{x^2 - 3x + 2} = \dfrac{2}{x - 2} + \dfrac{1}{x - 1} = \dfrac{2}{x^2 - 3x + 2} +$

$\dfrac{3}{x - 1} = \dfrac{-1}{x^2 - 3x + 2} + \dfrac{3}{x - 2}$

Chapter 5 Real numbers, fractional indices and surds

5.1 Operations with real numbers (1)

❶ C ❷ B ❸ C ❹ D ❺ $3\sqrt{2}$

❻ $2\sqrt{2} + \sqrt{3}$ ❼ $30\sqrt{7}$ ❽ $-2\sqrt{5}$ ❾ $\sqrt{5} - 3\sqrt{10}$ ❿ -3.05 ⓫ -5 ⓬ $-4 - 6\sqrt{6}$

⓭ (a) 16 (b) 4 [Hint: The answer is positive, and from (a) we get that its square is 16] ⓮ (a) 2018 (b) 4029 [Hint: $x^2 + 4028x + 2014 \times 2015 = (x + 2014)^2 + 2014$, and $x + 2014 = \sqrt{2015}$] ⓯ 9.80

5.2 Operations with real numbers (2)

❶ C ❷ B ❸ D ❹ 1.03 ❺ 0.41

❻ -1.38 ❼ 0.302 ❽ 3 ❾ $3\sqrt{2} + \sqrt{17}$ ❿ eg. $1 : \sqrt{3} = \sqrt{3} : 3$, $1 : 3 = \sqrt{3} : 3\sqrt{3}$,

$3 : 1 = \sqrt{3} : \dfrac{\sqrt{3}}{3}$ etc. ⓫ (a) $2\sqrt{5} - \sqrt{6} - \sqrt{2}$

(b) 5 ⓬ When $0 < a < 1$, it is $-2a + 3$; when $1 \leqslant a < 2$, it is 1 ⓭ 1.47 ⓮ (a) $9 - 2\sqrt{14}$ (b) 3

5.3 Fractional indices (1)

❶ D ❷ C ❸ B ❹ (a) $a^{\frac{p}{q}}$ (b) $a^{-\frac{p}{q}}$

❺ (a) $2^{\frac{3}{5}}$ (b) $5^{\frac{2}{3}}$ (c) $3^{-\frac{5}{2}}$ ❻ (a) $\sqrt[5]{3^4}$

(b) $\sqrt[3]{5^2}$ (c) $\dfrac{1}{\sqrt[5]{2^3}}$ ❼ 12 ❽ $2\dfrac{3}{16}$

❾ 6.538 ❿ 3.46 ⓫ (a) $5^{\frac{11}{12}}$ (b) $6^{\frac{7}{3}}$

(c) $7^{-\frac{5}{6}}$ ⑫ (a) $5^{\frac{17}{15}}$ (b) $6^{\frac{1}{6}}$ (c) $7^{-\frac{5}{6}}$

⑬ (a) $18-\sqrt{2}$ (b) $14\frac{1}{2}$ ⑭ $\frac{1}{16}$ [Hint: $x=8$, $y=-\frac{4}{3}$]

5.4 Fractional indices (2)

① B ② C ③ C ④ $\frac{1}{2}$ ⑤ $-\frac{4}{3}$

⑥ $2^{\frac{23}{12}}$ ⑦ $3^{\frac{5}{12}}$ ⑧ a^2-2+a^{-2} ⑨ $a^{\frac{1}{2}}+b^{\frac{1}{2}}$ ⑩ $a^{\frac{1}{2}}-2b^{\frac{1}{2}}$ [Hint: The original expression $=\dfrac{(a^{\frac{1}{2}}-b^{\frac{1}{2}})(a^{\frac{1}{2}}-2b^{\frac{1}{2}})}{a^{\frac{1}{2}}-b^{\frac{1}{2}}}=a^{\frac{1}{2}}-2b^{\frac{1}{2}}$] ⑪ 2 ⑫ (a) $2\sqrt{5}$ (b) $\sqrt{15}-3$

⑬ (a) 6 [Hint: $a+a^{-1}=(a^{\frac{1}{2}}-a^{-\frac{1}{2}})^2+2$]

(b) 34 ⑭ $2^{\frac{38}{15}}$ [Hint: $3^{4a+3b}=(3^a)^4\times(3^b)^3=(2^{\frac{1}{3}})^4\times(2^{\frac{2}{5}})^3=2^{\frac{38}{15}}$]

5.5 Surds (1)

① B ② D ③ $x\leqslant 0$ ④ $x\geqslant 0$ and $x\neq\frac{1}{2}$ ⑤ $\geqslant 0$ $\leqslant 0$ ⑥ $1-x$ ⑦ $4-2a$

⑧ ± 1 ⑨ 7 ⑩ $\sqrt{n+\dfrac{1}{n+2}}=(n+1)\sqrt{\dfrac{1}{n+2}}$ ⑪ (a) $\pi-3.14$ (b) -9 (c) $\frac{3}{2}$

(d) 36 ⑫ 0 ⑬ $3\sqrt{2}-2$ ⑭ 2 ⑮ 0 ⑯ 2, 3, 4 ⑰ 4

5.6 Surds (2)

① $x\geqslant 0$, $y\geqslant 0$ ② $x\geqslant 0$, $y>0$ ③ 4 7 4 7 $2\sqrt{7}$ ④ $x\geqslant 1$ ⑤ $x>2$

⑥ $\sqrt{27}$, $\sqrt{\dfrac{1}{2}}$ ⑦ $-m\sqrt{n}$ ⑧ (a) $2\sqrt{2}$

(b) $2\sqrt{19}$ (c) $\frac{4}{5}$ (d) $\frac{4}{3}$ (e) $4\sqrt{6}$

(f) $\frac{2b}{3a}$ ⑨ 8.8 ⑩ $\frac{9}{14}$ ⑪ $\frac{5y^2}{6x}$ ⑫ $\frac{\sqrt{3}}{1-a}$

⑬ $20x^2\sqrt{y}$ ⑭ $\frac{2}{5}$ ⑮ 1 ⑯ $-2b$

⑰ 15

5.7 Surds in simplest form

① $\sqrt{5ab}$, $\sqrt{a^2-b^2}$, $\frac{\sqrt{x}}{2}$ ② (a) $2\sqrt{3}$

(b) $3\sqrt{2x}$ (c) $\frac{\sqrt{xy}}{x}$ (d) $\frac{3\sqrt{2}}{2}$ ③ (a) $\sqrt{2}$

(b) $\sqrt{3a}$ (c) $\sqrt{3}$ (d) $\sqrt{3a}$ ④ $\frac{4\sqrt{5}}{25}$

⑤ $2ab\sqrt{10bc}$ ⑥ $\frac{2m\sqrt{15an}}{a}$ ⑦ $x\sqrt{x^2+3}$

⑧ $2\sqrt{x^4+4a^2}$ ⑨ $\frac{\sqrt{a^3+b^3}}{ab}$ ⑩ $2\sqrt{2}$

112 ⑪ $\sqrt{21}$ ⑫ $-\sqrt{2}$ ⑬ (a) $\sqrt{2}$

(b) $\sqrt{x-2y}$ (c) $\sqrt{x}+\sqrt{2y}$ (d) $2-\sqrt{3}$

(e) $3-2\sqrt{2}$ (f) $3\sqrt{2}+2\sqrt{3}$ ⑭ (a) $\sqrt{11}-\sqrt{10}$ $2\sqrt{2}-\sqrt{7}$ (b) $\sqrt{n+1}-\sqrt{n}$

5.8 Like quadratic surds

① $\frac{1}{3}\sqrt{12a^3x^3}$, $3a\sqrt{\dfrac{x}{3a}}$ ② $2\sqrt{45}$, $\frac{3}{5}\sqrt{125}$, $10\sqrt{0.05}$ ③ 4 ④ -15

⑤ $-4\sqrt{3}-2\sqrt{2}$ ⑥ $-\frac{14}{3}\sqrt{a}+\frac{21}{4}\sqrt{ab}$

⑦ $\frac{14}{5}\sqrt{x}+4y\sqrt{2x}$ ⑧ $3\sqrt{2}+7$ ⑨ $-\frac{2}{5}\sqrt{5}-\frac{2}{3}\sqrt{6}$ ⑩ $\frac{1}{2}\sqrt{abc}-\frac{17}{3}\sqrt{ab}$ ⑪ $3\sqrt{3}$

⑫ $\frac{25}{2}\sqrt{2}+\frac{5}{3}\sqrt{7}$ ⑬ $4\sqrt{2}-\frac{19}{5}\sqrt{3}$ ⑭ $x\sqrt{x}$

⑮ $\sqrt{ab}-4ab\sqrt{ab}$ ⑯ $a=1$, $b=4$

5.9 Adding and subtracting quadratic surds

① ①, ⑤, ⑥; ②, ⑦; ③, ④ ② ④

③ $-14\sqrt{2}$ ④ $\sqrt{3}-\sqrt{2}$ ⑤ $1-\sqrt{3}$

⑥ $15\sqrt{3}$ ⑦ $2\sqrt{3}+3\sqrt{5}$ ⑧ $\frac{\sqrt{5}}{5}$ ⑨ $\frac{11\sqrt{3}}{4}-\frac{1}{4}\sqrt{2}$ ⑩ $\sqrt{3}+\sqrt{2}$ ⑪ $14\sqrt{2x}$ ⑫ $3\sqrt{x}$

⑬ $\frac{1}{2}\sqrt{a}+3\sqrt{b}$ ⑭ 0 ⑮ $\frac{1}{2}\sqrt{x}+3\sqrt{y}$; 2

⑯ $-\sqrt{xy}$; $-\frac{9}{2}\sqrt{2}$. ⑰ The original expression

$$=\frac{(a-b)^2}{(a+b)(a-b)}\div\frac{b-a}{ab}=\frac{a-b}{a+b}\times\frac{ab}{b-a}=$$

$$\frac{-ab}{a+b} = \frac{-(\sqrt{2}+1)(\sqrt{2}-1)}{(\sqrt{2}+1)+(\sqrt{2}-1)} = -\frac{\sqrt{2}}{4}$$

⑱ From the given, we get $(2x-1)^2 + (y-3)^2 = 0$,

the solutions are $x = \frac{1}{2}$, $y = 3$. Therefore, the

original expression $= x\sqrt{x} + 6\sqrt{xy}$

$$= \frac{1}{2}\sqrt{\frac{1}{2}} + 6\sqrt{\frac{1}{2} \times 3} = \frac{\sqrt{2}}{4} + 3\sqrt{6}.$$

5.10 Multiplying and dividing quadratic surds

① $\sqrt{6}$ **②** 36 **③** 5 **④** $10\sqrt{2}a$ **⑤** $-\frac{4}{3}$

⑥ $18\sqrt{3}y$ **⑦** $-xy\sqrt{3x}$ **⑧** $6mn^2\sqrt{mn}$

⑨ 108 **⑩** $x^2y^2\sqrt{x}$ **⑪** $10\sqrt{2}$ **⑫** $-\frac{\sqrt{6}}{6}$

⑬ $-\frac{n^2}{m^4}\sqrt{mn}$ **⑭** $-\sqrt{6}a$ **⑮** From the given

we get $x = 2$ and $y = \frac{1}{4}$, then the original

expression $= \sqrt{2 + \frac{1}{4}} \times \sqrt{2 - \frac{1}{4}} = \frac{3\sqrt{7}}{4}$.   **⑯** $2\sqrt{2}$

⑰ From the given, we get $6 < x \le 9$ and x is an

even number, then $x = 8$. Therefore the original

expression $= (1+x)\sqrt{\frac{(x-4)(x-1)}{(x-1)(x+1)}}$

$$= (x+1)\sqrt{\frac{x-4}{x+1}} = \sqrt{(x-4)(x+1)}$$

$$= \sqrt{(8-4)(8+1)} = 6.$$

5.11 Rationalising denominators

① $\frac{\sqrt{2}}{4}$ **②** $\sqrt{2}+1$ **③** $\sqrt{3}-1$ **④** $-\sqrt{3}-2$

⑤ $\frac{\sqrt{2m}}{6m}$ **⑥** $\frac{\sqrt{x+y}}{x+y}$ **⑦** $\sqrt{x}-\sqrt{y}$

⑧ $\sqrt{7a}+\sqrt{5a}$ **⑨** $7-4\sqrt{3}$ **⑩** $\frac{\sqrt{x^2-y^2}}{x-y}$

⑪ $\sqrt{m}-\sqrt{n}$ **⑫** $x-\sqrt{x^2-1}$ **⑬** $2\sqrt{3}$

⑭ Since $\frac{1}{\sqrt{7}-\sqrt{5}} = \frac{\sqrt{7}+\sqrt{5}}{2}$, $\frac{1}{\sqrt{8}-\sqrt{6}} = \frac{\sqrt{8}+\sqrt{6}}{2}$

and $\frac{\sqrt{7}+\sqrt{5}}{2} - \frac{\sqrt{8}+\sqrt{6}}{2} < 0$, then $\frac{1}{\sqrt{7}-\sqrt{5}} <$

$\frac{1}{\sqrt{8}-\sqrt{6}}$.   **⑮** From the given, we get $a = -\sqrt{2}$

and $b = -1$, then the original expression $= \frac{1}{-1+\sqrt{2}}$

$= \sqrt{2}+1$.   **⑯** From the given, we get $x = 2 + \sqrt{3}$, then $x^2 - 4x + 2 = (x-2)^2 - 2 = 1$.

⑰ From the given, we get $a = 1004$ and $x = 5$,

then $\frac{\sqrt{x+1}-\sqrt{x}}{\sqrt{x+1}+\sqrt{x}} + \frac{\sqrt{x+1}+\sqrt{x}}{\sqrt{x+1}-\sqrt{x}}$

$$= \frac{(\sqrt{x+1}-\sqrt{x})^2 + (\sqrt{x+1}+\sqrt{x})^2}{x+1-x} = 4x+2$$

$= 22.$ **⑱** The original expression $= (2\sqrt{5}+1)$

$\left(\frac{\sqrt{2}-1}{2-1} + \frac{\sqrt{3}-\sqrt{2}}{3-2} + \frac{\sqrt{4}-\sqrt{3}}{4-3} + \cdots + \frac{\sqrt{100}-\sqrt{99}}{100-99} \right)$

$= (2\sqrt{5}+1)[(\sqrt{2}-1) + (\sqrt{3}-\sqrt{2}) + (\sqrt{4}-\sqrt{3})$

$+ \cdots + (\sqrt{100}-\sqrt{99})] = (2\sqrt{5}+1)(\sqrt{100} - 1) = 9(2\sqrt{5}+1).$

5.12 Mixed operations with quadratic surds

① $2\sqrt{2}$ **②** $-3\sqrt{ax}$ **③** $2\sqrt{7}$ **④** $6+3\sqrt{3}-2\sqrt{2}-\sqrt{6}$ **⑤** $-\frac{1}{4}$ **⑥** $-18-2\sqrt{6}$ **⑦** $\frac{\sqrt{6}}{6}$

⑧ $84-24\sqrt{6}$ **⑨** $\frac{17}{2} - \frac{14}{3}\sqrt{3}$ **⑩** $\frac{29}{4}$

⑪ $15\sqrt{2}$ **⑫** 2 **⑬** $1-\sqrt{2}$ **⑭** $4\sqrt{ab}$

⑮ (a) $x^2 - xy + y^2 = (x-y)^2 + xy = 8+1 = 9$

(b) $x^3y + xy^3 = xy(x^2+y^2) = 10$ **⑯** 4

⑰ From the given, we get $a = 3$ and $b = \sqrt{3}-1$,

then the original expression $= (a+b)^2 - ab =$

$(3+\sqrt{3}-1)^2 - 3(\sqrt{3}-1) = 10 + \sqrt{3}$

Unit test 5

① B   **②** C   **③** D   **④** D   **⑤** B   **⑥** A

⑦ B   **⑧** <0 **⑨** $a \ge 3$ **⑩** $3\frac{2}{3}$

⑪ $b-a$ **⑫** $-2\sqrt{mn}$ **⑬** $\frac{19\sqrt{2a}}{4}$ **⑭** 6

⑮ $69-18\sqrt{10}$ **⑯** 1 **⑰** $\pm\frac{1}{3}$ **⑱** $-\frac{1}{a}$

⑲ 1 **⑳** $5\sqrt{2}+2\sqrt{3}$ **㉑** $\frac{2}{9}$ **㉒** $3\sqrt{x}$

㉓ $5 + 4\sqrt{3}$ ㉔ $\dfrac{-9\,a^2\,\sqrt{ab}}{b}$ ㉕ $-2 + \sqrt{2} +$

$3\sqrt{3} + \sqrt{6}$ ㉖ $\dfrac{\sqrt{6}}{2}$ ㉗ $9 + 2\sqrt{10}$ ㉘ $\dfrac{2\sqrt{3}}{3}$

㉙ (a) $\sqrt{9999 \times 9999 + 19\,999} = 10\,000$.

Proof: $\sqrt{9999 \times 9999 + 19\,999} =$

$\sqrt{9999 \times 9999 + 10\,000 + 9999} =$

$\sqrt{9999 \times (9999 + 1) + 10\,000} =$

$\sqrt{9999 \times 10\,000 + 10\,000} =$

$\sqrt{10\,000 \cdot (9999 + 1)} = 10000$ (b) The nth

equation: $\sqrt{\underbrace{999\cdots999}_{n} \times \underbrace{999\cdots999}_{n} + 1\underbrace{999\cdots999}_{n}}$

$= 1\underbrace{000\cdots000}_{n}$ ㉚ (a) $\sqrt{7 + 4\sqrt{3}} =$

$\sqrt{(\sqrt{4})^2 + 2\sqrt{4 \times 3} + (\sqrt{3})^2} = \sqrt{(\sqrt{4} + \sqrt{3})^2} =$

$2 + \sqrt{3}$ (b) $\sqrt{13 - 2\sqrt{42}} =$

$\sqrt{(\sqrt{7})^2 - 2\sqrt{6 \times 7} + (\sqrt{6})^2} = \sqrt{(\sqrt{7} - \sqrt{6})^2} =$

$\sqrt{7} - \sqrt{6}$

Chapter 6 Quadratic equations

6.1 Concepts of quadratic equations

① A ② B ③ 5 -6 8 ④ 13

⑤ -1 ⑥ $m \neq 0$ ⑦ $\neq 3$ ⑧ $x^2 + x +$

$5 = 0$ ⑨ 3, -1, 0 ⑩ 1 ⑪ (a) 1, $-\dfrac{1}{3}$

(b) $\dfrac{1}{2}$, 1 ⑫ (a) $k \neq 2$ (b) $k = 2$

(c) Yes, because $x = -1$ makes the value on both

sides of the equation equal. ⑬ When $m = -2$,

the original equation is a quadratic equation in one

variable. When $m = 0$ or -1 or 2, the original

equation is a linear equation in one variable.

⑭ (a) $m \neq \dfrac{1}{2}$ (b) The value of m can be any

number.

6.2 Solving quadratic equations (1): by taking square roots

① B ② C ③ $x = \pm 2$ ④ $\pm\sqrt{2}$

⑤ has no real roots ⑥ $b \leqslant 0$ ⑦ $a \neq 0$, a

and b are of the same sign. Or $b = 0$ ⑧ 8

⑨ (a) $x = \pm 7$ (b) $x = \pm 3$ (c) Since $y^2 =$

$\dfrac{5}{3}$, $y = \pm\dfrac{\sqrt{15}}{3}$ (d) Since $x - \sqrt{5} = \pm 2$, $x =$

$\sqrt{5} \pm 2$.   (e) From $(2x + 1)^2 = 32$ we get $2x +$

$1 = \pm 4\sqrt{2}$, therefore $x = -\dfrac{1}{2} \pm 2\sqrt{2}$.   (f) From

the given we get $(x + 6)^2 = 27$, then $x + 6 =$

$\pm 3\sqrt{3}$, therefore $x = -6 \pm 3\sqrt{3}$.   (g) From the

original equation we get $2x + 1 = \pm (3x - 2)$, then

the solutions are: $x_1 = 3$, $x_2 = \dfrac{1}{5}$.   (h) From

the original equation we get $2x - 1 = \pm (x + 1)$,

then the solutions are: $x_1 = 2$, $x_2 = 0$ (i) From

the equation we get $x - a = \pm (a + b)$, then $x_1 =$

$2a + b$, $x_2 = -b$.   (j) When $n > 0$, $x = m \pm$

$\sqrt{n}$. When $n = 0$, $x_1 = x_2 = m$. When $n < 0$, there

is no real roots in the original equation.

⑩ From the given we get when $a > 0$, $b = 0$ and

$c < 0$, then $a > b > c$.

6.3 Solving quadratic equations (2): by factorising

① C ② D ③ $x(x - 5)$ ④ $(x + 12)(x +$

8) $x_1 = -12$, $x_2 = -8$ ⑤ $x_1 = -7$, $x_2 = 1$

⑥ From the original equation, we get: $2(2x -$

$1)(x - 1) = 0$, then the solutions are: $x_1 = \dfrac{1}{2}$,

$x_2 = 1$.   ⑦ Substituting $x = 0$ into the equation

gives: $m^2 + 5m + 6 = 0$, that is: $(m + 2)(m + 3) =$

0, when $m + 3 \neq 0$, then $m = -2$.   ⑧ 0, 2 or

-2 [Hint: Since $x^2 = (\sqrt{x^2})^2$ is true, $(\sqrt{x^2})^2 -$

$2\sqrt{x^2} = 0$, that is: $\sqrt{x^2}(\sqrt{x^2} - 2) = 0$. Therefore

$\sqrt{x^2} = 0$, $\sqrt{x^2} = 2$. The solutions are: $x_1 = 0$,

$x_2 = 2$, $x_3 = -2$]   ⑨ (a) Simplifying $(x +$

$3)(x - 6) = 0$ gives $x^2 - 3x - 18 = 0$

(b) Simplifying $[x - (\sqrt{5} - 4)][x - (\sqrt{5} + 4)] =$

0 gives $x^2 - 2\sqrt{5}x - 11 = 0$. To make the

coefficient of the linear term 2, we get $-\dfrac{\sqrt{5}}{5}x^2 + 2x$

$+ \dfrac{11\sqrt{5}}{5} = 0$.   ⑩ (a) Factorising the original

equation we get $x(5x - 4) = 0$, therefore $x_1 = 0$, $x_2 = \dfrac{4}{5}$. (b) $x_1 = \dfrac{5}{2}$, $x_2 = -\dfrac{5}{2}$ (c) Factorising the original equation we get $(x - 2)(1 - x) = 0$, therefore $x_1 = 1$, $x_2 = 2$. (d) Factorising the original equation we get $[(x + 1) + 5][(x + 1) - 5] = 0$, $x + 6 = 0$ or $x - 4 = 0$, therefore $x_1 = -6$, $x_2 = 4$. (e) From the original equation we get $y^2 + 5y - 24 = 0$, and then factorising the equation $(y + 8)(y - 3) = 0$ gives $y_1 = -8$, $y_2 = 3$.
(f) Factorising the original equation $(3x + 3)(-x + 5) = 0$ gives $x_1 = -1$, $x_2 = 5$ (g) From the original equation we get $(3 - y)^2 + y^2 - 9 = 0$, and then factorising the equation $2y(y - 3) = 0$ gives $y_1 = 0$, $y_2 = 3$. (h) Factorising the original equation $(2x - \sqrt{3})(x + \sqrt{3} + 1) = 0$ gives $x_1 = \dfrac{\sqrt{3}}{2}$, $x_2 = -\sqrt{3} - 1$. ⑪ The two roots of the equation $x^2 - 6x + 8 = 0$ are $x_1 = 2$ and $x_2 = 4$. If the third side of the isosceles triangle is 2, this could not be a triangle. Therefore, the third side is 4. The perimeter of this triangle is 10.

⑫ Factorising the original equation gives $(x - a)[(a - 1)x - (a + 1)] = 0$. When $a - 1 \neq 0$, that is $a \neq 1$, $x_1 = a$, $x_2 = \dfrac{a + 1}{a - 1} = 1 + \dfrac{2}{a - 1}$. Since the roots of the original equation are positive integers, given $a - 1 = 1$ or $a - 1 = 2$, $a = 2$ or 3. If $a - 1 = 0$, then when $a = 1$, the solution to the original equation is $x = 1$, which satisfies the condition. From the above, the integer value of a is $a = 1$ or $a = 2$ or $a = 3$.

6.4 Solving quadratic equations (3): by completing the square

① C ② B ③ 16 4 ④ $9m^2$ 3m
⑤ $(a + b)^2$ $(a + b)$ ⑥ $(x - 1)^2 - 3$
⑦ ±3 ⑧ $3\left[\left(x + \dfrac{\sqrt{2}}{6}\right)^2 - \dfrac{37}{18}\right] = 0$ ⑨ 1
[Hint: by completing the square $(x + 2)^2 + (y - 3)^2 = 0$, $x = -2$, $y = 3$. Therefore $x + y = 1$]
⑩ positive ⑪ (a) Completing the square

$(x - 1)^2 = 6$ gives $x = 1 \pm \sqrt{6}$. (b) Completing the square $(x - 2)^2 = 10\,000$ gives $x_1 = 102$, $x_2 = -98$. (c) Completing the square $\left(x - \dfrac{3}{10}\right)^2 = \dfrac{1}{4}$ gives $x_1 = \dfrac{4}{5}$, $x_2 = -\dfrac{1}{5}$.
(d) Completing the square $\left(x + \dfrac{3}{4}\right)^2 = \dfrac{65}{16}$ gives $x = -\dfrac{3}{4} \pm \dfrac{\sqrt{65}}{4}$. (e) Completing the square $\left(x + \dfrac{1}{12}\right)^2 = \dfrac{49}{144}$ gives $x_1 = \dfrac{1}{2}$, $x_2 = -\dfrac{2}{3}$.
(f) Completing the square $\left(y + \dfrac{1}{4}\right)^2 = \dfrac{49}{16}$ gives $y_1 = \dfrac{3}{2}$, $y_2 = -2$. (g) Completing the square $(x + 2)^2 = 8$ gives $x = -2 \pm 2\sqrt{2}$. (h) Completing the square $\left(x - \dfrac{2}{3}\right)^2 = \dfrac{10}{9}$ gives $x = \dfrac{2}{3} \pm \dfrac{\sqrt{10}}{3}$.
(i) Completing the square $\left(x - \dfrac{5a}{2}\right)^2 = \dfrac{49 a^2}{4}$ gives $x_1 = 6a$, $x_2 = -a$. (j) Completing the square $(y + \sqrt{3} + 1)^2 = 4$ gives $y_1 = 1 - \sqrt{3}$, $y_2 = -3 - \sqrt{3}$ ⑫ From $x^2 + 6x + y^2 - 8y + 25 = 0$ we get $(x + 3)^2 + (y - 4)^2 = 0$, then $x = -3$, $y = 4$. Therefore $\dfrac{x - 2y}{x^2 + y^2} = -\dfrac{11}{25}$.

6.5 Solving quadratic equations (4): by using the quadratic formula

① A ② B ③ $x = \dfrac{-b \pm \sqrt{b^2 - 4ac}}{2a}$
④ 9 ⑤ 36 ⑥ $-\dfrac{2}{3}$, 1 ⑦ (a) Since $\Delta = 9 + 16 = 25$, $x = \dfrac{3 \pm \sqrt{25}}{2}$. Therefore $x_1 = 4$, $x_2 = -1$. (b) From the original equation we get $x^2 - 2x + 2 = 0$, then $\Delta = 4 - 8 = -4 < 0$. Therefore there are no real roots in the original equation. (c) Since $\Delta = 48 - 40 = 8$, $x = \dfrac{4\sqrt{3} \pm \sqrt{8}}{2} = 2\sqrt{3} \pm \sqrt{2}$. (d) From the original equation we get $5x^2 + 3x - 2 = 0$, $\Delta = 9 + 40 = 49$,

then $x = \dfrac{-3 \pm \sqrt{49}}{10}$. Therefore $x_1 = \dfrac{2}{5}$, $x_2 = $ -1. (e) Since $\Delta = (-2)^2 - 4 \times 5 \times (-7) = $ 144, $x = \dfrac{2 \pm 12}{10}$. Therefore $x_1 = \dfrac{7}{5}$, $x_2 = -1$.

(f) From the original equation we get $4x^2 - x + 1 = 0$, then $\Delta = (-1)^2 - 4 \times 4 \times 1 = -15 < 0$. Therefore there are no real roots in the original equation. (g) Since $\Delta = (2\sqrt{3})^2 - 4 \times 1 \times 3 = 0$, $x_1 = x_2 = -\sqrt{3}$. (h) From the original equation we get $x^2 + 4x - 2 = 0$, then $\Delta = 4^2 - 4 \times 1 \times (-2) = 24$. Therefore $x = \dfrac{-4 \pm \sqrt{24}}{2 \times 1} = \dfrac{-4 \pm 2\sqrt{6}}{2} = -2 \pm \sqrt{6}$. (i) Since $\Delta = 28 - 8 = 20$, $x = \dfrac{-2\sqrt{7} \pm \sqrt{20}}{2 \times 2} = \dfrac{-2\sqrt{7} \pm 2\sqrt{5}}{4} = \dfrac{-\sqrt{7} \pm \sqrt{5}}{2}$. (j) From the original equation we get $x^2 - 2\sqrt{2}x - 1 = 0$, then $\Delta = 8 + 4 = 12$. Therefore $x = \dfrac{2\sqrt{2} \pm \sqrt{12}}{2} = \sqrt{2} \pm \sqrt{3}$.

8 (a) From the original equation we get $x^2 - x - a^2 = 0$, then $\Delta = 1 + 4a^2$. Therefore $x = \dfrac{1 \pm \sqrt{1 + 4a^2}}{2}$. (b) Since $\Delta = (11mn)^2 - 4 \times (20m^2) \times (-3n^2) = (19mn)^2$, $x = \dfrac{-11mn \pm 19mn}{2 \times 20\,m^2}$. Therefore $x_1 = \dfrac{n}{5m}$, $x_2 = -\dfrac{3n}{4m}$.

6.6 Solving quadratic equations (5): by suitable methods

1 C **2** C **3** taking the square roots 3 -1 **4** factorising 2 1 **5** completing the square 21 -19 **6** -3 or $\dfrac{3}{2}$ **7** 2

8 $x = -1 \pm \sqrt{1 - c}$ **9** (a) $x_1 = \dfrac{1}{3}$, $x_2 = -1$

(b) $x_1 = 10$, $x_2 = -6$ (c) $x_1 = 3$, $x_2 = \dfrac{14}{5}$

(d) $x_1 = \dfrac{7}{5}$, $x_2 = -1$ (e) No real roots

(f) $x_1 = x_2 = -\dfrac{\sqrt{3}}{3}$ (g) $x_1 = 2$, $x_2 = 3$

(h) $x_1 = 8$, $x_2 = \dfrac{2}{3}$ (i) $x = -\sqrt{6} \pm 2\sqrt{2}$

(j) $x_1 = -3(\sqrt{2} + 1)$, $x_2 = \sqrt{2} + 1$ (k) $x_1 = \dfrac{bc}{a}$, $x_2 = b + c$ (l) When $m = 0$, $x = -2$; when $m \neq 0$, $x_1 = -\dfrac{2m + 1}{m}$, $x_2 = -2$. **10** When the value of y is equal to the value of $4x + 1$, then $2x^2 + 7x - 1 = 4x + 1$, $2x^2 + 3x - 2 = 0$. The solutions are: $x_1 = -2$, $x_2 = \dfrac{1}{2}$. When the value of y and the value of $x^2 - 19$ added to zero, then $(2x^2 + 7x - 1) + (x^2 - 19) = 0$, $3x^2 + 7x - 20 = 0$. The solutions are: $x_1 = -4$, $x_2 = \dfrac{5}{3}$. **11** Since a is one of the roots, $a^2 - a - 1 = 0$, $a^2 = a + 1$. Therefore $a^3 - 2a + 3 = a^2 \cdot a - 2a + 3 = (a + 1)a - 2a + 3 = a^2 - a + 3 = 1 + 3 = 4$ **12** Since $x > 0$ and $x^2 = 1 - x$, $x = \dfrac{1}{x} - 1$, so $x - \dfrac{1}{x} = -1$.

Therefore $\left(x - \dfrac{1}{x}\right)^2 = 1$, $x^2 + \dfrac{1}{x^2} = 1 + 2 = 3$.

Since $\left(x + \dfrac{1}{x}\right)^2 = x^2 + \dfrac{1}{x^2} + 2 = 3 + 2 = 5$ and $x + \dfrac{1}{x} > 0$, $x + \dfrac{1}{x} = \sqrt{5}$.

6.7 Discriminant of a quadratic equation (1)

1 D **2** B **3** $b^2 - 4ac > 0$, $= 0$, no, $\dfrac{-b + \sqrt{b^2 - 4ac}}{2a}$, $\dfrac{-b - \sqrt{b^2 - 4ac}}{2a}$ **4** 24

5 $<$ no **6** 3 two distinct real **7** 4

8 2 **9** (a) Since $\Delta = 41 > 0$, there are two distinct real roots. (b) Since $\Delta = -8 < 0$, there are no real roots. (c) Since $\Delta = 0$, there are two equal real roots. (d) Since $\Delta = 64 > 0$, there are two distinct real roots. (e) Since $\Delta = 1 + 4\sqrt{6} > 0$, there are two distinct real roots. (f) Since $\Delta = -16 < 0$, there are no real roots. **10** (a) Since $a \neq 0$, it is a quadratic equation and $\Delta = (-b)^2 - 4 \cdot a \cdot 0 = b^2 \geqslant 0$, therefore there are two real roots. (b) Since $\Delta = -12k^2 - 16 < 0$, there are no real roots. (c) Since $\Delta = -m^2 - 4m - 6 = -(m + 2)^2 - 2 < 0$, there are no real roots. (d) Since $\Delta = 8ab^2c - 4a^2c^2 - 4b^4$ and $b^2 = ac$, $\Delta = 0$. Therefore there are two equal real roots. ⑪ Since $\Delta = 4m^2 - 16m + 16 = 4(m-2)^2 \geqslant 0$, it definitely has two real roots.

⑫ Since there are no real roots in the equation $x^2 - 2x - m = 0$, $\Delta_1 = 4 + 4m < 0$, $m < -1$ and $\Delta_2 = -4m > 0$. Therefore there are two distinct real roots in the equation $x^2 + 2mx + m(m + 1) = 0$.

6.8 Discriminant of a quadratic equation (2)

① A ② B ③ –1 or 2 ④ $m \leqslant 1$ and $m \neq 0$ ⑤ $m > -\dfrac{1}{4}$ ⑥ $k < -\dfrac{9}{8}$ ⑦ $k \leqslant \dfrac{9}{8}$ and $k \neq 0$ ⑧ 1 ⑨ From the given $\Delta = (-4)^2 - 4(k - 5) = 36 - 4k$. (a) Since there are two distinct real roots in the equation, $\Delta > 0$ and $k < 9$. (b) Since there are two equal real roots in the equation, $\Delta = 0$ and $k = 9$. (c) Since there are no real roots in the equation, $\Delta < 0$ and $k > 9$. ⑩ From the given we get $\Delta = (-2m)^2 - 4(m^2 + 1)(m^2 + 4) = -4(m^4 + 4m^2 + 4) = -4(m^2 + 2)^2$. Since the set of m can be any real number $(m^2 + 2)^2 > 0$, $-4(m^2 + 2)^2 < 0$, that is $\Delta < 0$. Therefore the equation $(m^2 + 1)x^2 - 2mx + (m^2 + 4) = 0$ has no real roots. ⑪ (1) If $k^2 - 1 = 0$, then $k = \pm 1$. When $k = 1$, the root of the equation is $x = -\dfrac{1}{4}$; when $k = -1$, there are no solutions to the equation. (2) If $k^2 - 1 \neq 0$, then $k \neq \pm 1$, it is a quadratic equation. If the equation has real roots and $\Delta \geqslant 0$, then $k \geqslant -1$. Therefore $k > -1$ and $k \neq 1$. Thus when $k > -1$, there are real roots in the equation. ⑫ (a) From $\Delta > 0$ we get $k < 2$. Since $2k + 4 \geqslant 0$, $k \geqslant -2$. Therefore the set of possible value of k is $-2 \leqslant k < 2$. (b) Since $-2 \leqslant k < 2$, $\sqrt{k^2 + 4k + 4} +$

$\sqrt{k^2 - 4k + 4} = \sqrt{(k + 2)^2} + \sqrt{(k - 2)^2} = k + 2 + 2 - k = 4$ ⑬ Since there are two real roots in the given equation, $\Delta = 0$, that is $[\sqrt{2}(a - c)]^2 - 4(b + c)\left[-\dfrac{3}{4}(a - c)\right] = 0$ and $(a - c)(2a + 3b + c) = 0$. Since a, b and c are the three sides in $\triangle ABC$ and $a > 0$, $b > 0$, $c > 0$, $2a + 3b + c \neq 0$, $a - c = 0$ and $a = c$. Therefore $\triangle ABC$ is an isosceles triangle.

6.9 Applications of quadratic equations (1): factorising quadratic expressions

① C ② C ③ $3\left(x - \dfrac{-2 + \sqrt{7}}{3}\right)\left(x + \dfrac{2 + \sqrt{7}}{3}\right)$ ④ $2y - \dfrac{2}{3}y$ $3(x - 2y)\left(x + \dfrac{2}{3}y\right)$ ⑤ $-2\left(x + \dfrac{3 + \sqrt{57}}{4}\right)\left(x + \dfrac{3 - \sqrt{57}}{4}\right)$

⑥ 10 ⑦ $m \leqslant \dfrac{1}{8}$ ⑧ $k \leqslant \dfrac{1}{24}$

⑨ (a) $(x - 2)(x - 3)$ (b) $(2x + \sqrt{5})(2x - \sqrt{5})$ (c) $(2x + 2 - \sqrt{5})(2x + 2 + \sqrt{5})$

(d) $3\left(t - \dfrac{2 + \sqrt{7}}{3}\right)\left(t - \dfrac{2 - \sqrt{7}}{3}\right)$

(e) $2\left(y - \dfrac{2 + \sqrt{2}}{2}\right)\left(y - \dfrac{2 - \sqrt{2}}{2}\right)$

(f) $2\left(x - \dfrac{4 + \sqrt{6}}{2}y\right)\left(x - \dfrac{4 - \sqrt{6}}{2}y\right)$

(g) $3\left(xy - \dfrac{5 + \sqrt{37}}{6}\right)\left(xy - \dfrac{5 - \sqrt{37}}{6}\right)$

(h) $-4\left(x + \dfrac{2 + \sqrt{5}}{2}y\right)\left(x + \dfrac{2 - \sqrt{5}}{2}y\right)$

(i) $2x\left(x - \dfrac{2 + \sqrt{6}}{2}y\right)\left(x - \dfrac{2 - \sqrt{6}}{2}y\right)$

(j) $\left(x - \dfrac{-1 + \sqrt{13}}{2}\right)\left(x - \dfrac{-1 - \sqrt{13}}{2}\right)(x^2 + x + 1)$ ⑩ Since $\Delta = 4(2a + b)(2a - b)$ and a is the length of its two equal sides and b is the length of its base, both $2a + b$ and $2a - b$ are greater than 0, thus $\Delta > 0$. Therefore, the quadratic trinomial in x, $x^2 - 4ax + b^2$, must be factorised over real numbers. ⑪ From the given we get $4x^2 - kx +$

$1 = (2x - \sqrt{2} + 1)(tx + m)$. Compare the coefficients of x^2 on both sides and we get $4 = 2t$, so $t = 2$; compare the constant terms and we get $m(1 - \sqrt{2}) = 1$, so $m = -(1 + \sqrt{2})$; compare the coefficients of the terms x in both sides and we get $-k = 2m + t(1 - \sqrt{2}) = -2 - 2\sqrt{2} + 2 - 2\sqrt{2} = -4\sqrt{2}$. Therefore, $k = 4\sqrt{2}$.

6.10 Applications of quadratic equations (2): practical applications

1 B **2** C **3** $\frac{5}{2}$, 8 or 4, 5 **4** 10

5 $[1000(1 + x) - 200](1 + x) = 892.5$

6 $10(x + 2) + x = 3x^2$ **7** Let number of teams of football league be x, so the equation is: $x(x - 1) = 182$ and the solutions are: $x_1 = 14$, $x_2 = -13$ (reject). There are 14 teams in a football league. **8** The annual revenue of the company in 2017: $6 \div 40\% = 15$ million. Let the annual growth rate from 2017 to 2019 be x. From the company's estimation £21.6 million for year 2019, we get $15(1 + x)^2 = 21.6$, $(1 + x)^2 = 1.44$, that is: $1 + x = \pm 1.2$ ($1 + x = -1.2$ reject). Therefore $15(1 + x) = 15 \times 1.2 = £18$ million. The annual revenue expected in 2018 is £18 million. **9** Let the width of the path be x metres. The length and width of the 6 fields are: $\frac{1}{3}(32 - 2x)$ and $\frac{1}{2}(20 - x)$, so the equation is: $6 \times \frac{1}{3}(32 - 2x) \times \frac{1}{2}(20 - x) = 504$ and the solutions are: $x_1 = 2$, $x_2 = 34$ (reject). The width of the path is 2 metres. **10** Let the side length perpendicular to the wall be x metres, so the equation is: $x(32 + 1 - 2x) = 130$ and the solutions are: $x_1 = \frac{13}{2}$, $x_2 = 10$. When $x_1 = \frac{13}{2}$, $32 + 1 - 2x = 20 > 16$, $x_1 = \frac{13}{2}$ (reject). The length and width of the shed are 13 m and 10 m.

Unit test 6

1 D **2** A **3** B **4** B **5** D

6 9, 3 or 8, 7 **7** −6 **8** 20 **9** −2

10 1 or $-\frac{2}{3}$ **11** 4 **12** 1 −2 **13** −6

$3 + \sqrt{2}$ **14** $\pm 2\sqrt{3}$ **15** 20 **16** $x_1 = 0$, $x_2 = \frac{1}{2}$ **17** $m = 1$, $n = \frac{1}{4}$ (answer may vary. It can be any pair of m and n so that $m^2 = 4n$.)

18 $3\left(x + \frac{6 + \sqrt{3}}{3}y\right)\left(x + \frac{6 - \sqrt{3}}{3}y\right)$ **19** (a) $x_1 = 1$, $x_2 = 2$ (b) $x_1 = \frac{16}{3}$, $x_2 = \frac{4}{7}$ (c) $x_1 = \frac{1}{2}$, $x_2 = -2$ (d) $x_1 = 1$, $x_2 = 1 + k$

20 (a) $m < 1$ and $m \neq -1$ (b) It is impossible for the equation to have two equal real roots. (c) $m > 1$ **21** (a) Since equation $\frac{1}{2}x^2 + \sqrt{b}x + c - \frac{1}{2}a = 0$ in x has two equal real roots, $\Delta = (\sqrt{b})^2 - 4 \times \frac{1}{2}\left(c - \frac{1}{2}a\right) = 0$. Therefore $a + b - 2c = 0$. ① Since the root of the equation $3cx + 2b = 2a$ is $x = 0$, then $a = b$ ② Substitute ① into ②, we get $a = c$, so $a = b = c$. Therefore $\triangle ABC$ is an equilateral triangle. (b) Since a and b are the two roots of equation $x^2 + mx - 3m = 0$, then $m^2 - 4 \times (-3m) = 0$, that is: $m^2 + 12m = 0$. Then $m_1 = 0$, $m_2 = -12$. When $m = 0$, the solution is: $x = 0$ (not applicable; reject). Therefore $m = -12$. **22** Yes, there are mistakes in the solution. (a) If there are two distinct real roots of the equation, then it is a quadratic equation, so there must be $a^2 \neq 0$ and $(2a - 1)^2 - 4a^2 > 0$. Therefore, the correct answer is $a < \frac{1}{4}$ and $a \neq 0$.

(b) It is impossible that a equals $\frac{1}{2}$ because from (a), we know when the equation has two distinct real roots, the set of values that a can take is $a < \frac{1}{4}$ and $a \neq 0$, but $a = \frac{1}{2} > \frac{1}{4}$. In other words, when $a = \frac{1}{2}$, the quadratic equation has no real roots. Therefore, there is no such a real number a satisfying the conditions. **㉓** Let the percentage be x. So the equation is: $200 + 200(1 + x) + 200(1 + x)^2 = 1400$ and the solutions are: $x_1 = 1$, $x_2 = -4$ (reject). The growth rate is 100%.

㉔ (a) 2 seconds or 4 seconds (b) The area of pentagon $APQCD$ is the smallest after 3 seconds. It is 63 cm^2.

Chapter 7 Quadratic equations in two variables

7.1 Introduction to quadratic equations in two variables

① C **②** A **③** D **④** two; two

⑤ $-5x^2$, $-3xy$, $4y^2$; $-\dfrac{1}{2}$, 1; -1 **⑥** $xy = 4$ (answer may vary) **⑦** $x + y = 3$, $x + y = -3$ **⑧** $x - 2y = 0$, $x - y = 0$ **⑨** $x - y + 1 = 0$, $x - y - 2 = 0$ **⑩** $x - 2y - 7 = 0$, $x - 2y + 4 = 0$ **⑪** $2x - 3y = 0$, $x + y = 0$ **⑫** $\begin{cases} x = 0, \\ y = -\dfrac{1}{2} \end{cases}$ $\begin{cases} x = 1, \\ y = 0 \end{cases}$ $\begin{cases} x = 2, \\ y = \dfrac{3}{2} \end{cases}$ (answer may vary)

⑬ $\begin{cases} x_1 = 2, \\ y_1 = 3 \end{cases}$ $\begin{cases} x_2 = 2, \\ y_2 = -3 \end{cases}$ $\begin{cases} x_3 = -2, \\ y_3 = 3 \end{cases}$ $\begin{cases} x_4 = -2, \\ y_4 = -3 \end{cases}$

⑭ $\begin{cases} x_1 = 2, \\ y_1 = 1 \end{cases}$ $\begin{cases} x_2 = 0, \\ y_2 = -1 \end{cases}$

7.2 Solving simultaneous quadratic equations in two variables

① $\begin{cases} x_1 = 2, \\ y_1 = 3 \end{cases}$ $\begin{cases} x_2 = 3, \\ y_2 = 2 \end{cases}$ **②** $\begin{cases} x_1 = 2, \\ y_1 = 3 \end{cases}$ $\begin{cases} x_2 = -1, \\ y_2 = 0 \end{cases}$ **③** $\begin{cases} x_1 = -\dfrac{3}{2}, \\ y_1 = -4 \end{cases}$ $\begin{cases} x_2 = 4, \\ y_2 = 7 \end{cases}$

④ $\begin{cases} x_1 = -2, \\ y_1 = 7 \end{cases}$ $\begin{cases} x_2 = 7, \\ y_2 = -2 \end{cases}$ **⑤** $\begin{cases} x_1 = -2, \\ y_1 = -3 \end{cases}$ $\begin{cases} x_2 = 2, \\ y_2 = 1 \end{cases}$ **⑥** $\begin{cases} x_1 = 5, \\ y_1 = -\dfrac{1}{5} \end{cases}$ $\begin{cases} x_2 = -1, \\ y_2 = 1 \end{cases}$

⑦ $\begin{cases} x_1 = 0, \\ y_1 = -1 \end{cases}$ $\begin{cases} x_2 = -\dfrac{1}{2}, \\ y_2 = -2 \end{cases}$ **⑧** $\begin{cases} x_1 = -3, \\ y_1 = -4 \end{cases}$ $\begin{cases} x_2 = 2, \\ y_2 = 1 \end{cases}$ **⑨** $\begin{cases} x_1 = 3, \\ y_1 = 4 \end{cases}$ $\begin{cases} x_2 = 4, \\ y_2 = 3 \end{cases}$

⑩ $m = \pm 2\sqrt{10}$. When $m = 2\sqrt{10}$, $\begin{cases} x = \sqrt{10}, \\ y = \sqrt{10} \end{cases}$; when $m = -2\sqrt{10}$, $\begin{cases} x = -\sqrt{10}, \\ y = -\sqrt{10} \end{cases}$ **⑪** (a) $n < \dfrac{1}{2}$ (b) When $n = 0$, $\begin{cases} x_1 = 0, \\ y_1 = 0 \end{cases}$ and $\begin{cases} x_2 = 1, \\ y_2 = 2 \end{cases}$

7.3 Using quadratic (simultaneous) equations to solve application problems (1)

① Let the average monthly growth rate of the revenues from March to May be x. From the question we get $4(1 + 10\%)(1 + x)^2 = 6.336$, so $x = 20\%$ (reject the negative answer). The monthly growth rate of the revenue from March to May is 20% **②** From the question we get $(a - 21)(350 - 10a) = 400$, so $a_1 = 31$, $a_2 = 25$, since $a_1 - 21 > 4.2$, then reject a_1. Therefore, the selling price should be set £25 per pair. The shop should purchase 100 pairs.

③ Let the annual interest rate of the savings for the first time be x. From the question we get $[1000(1 + x) - 500](1 + 0.9x) = 530$, so $x = 0.0204 = 2.04\%$ (the negative value is not applicable, reject). The annual interest rate of the savings for the first time is 2.04% **④** Let the two non-hypotenuse sides of the triangle be a cm and b cm, then $\begin{cases} a + b = 17, \\ a^2 + b^2 = 169 \end{cases}$, the solutions are $\begin{cases} a = 5, \\ b = 12 \end{cases}$ or $\begin{cases} a = 12, \\ b = 5 \end{cases}$. Therefore the area of the right-angled triangle is 30 cm^2. **⑤** Let the cost price be x and the profit rate be y. Selling price – cost price = cost price $\times$ profit rate, i.e., $20 - x = xy$ and $0.5x + 1.5y = 20$, therefore $x = 10$

7.4 Using quadratic (simultaneous) equations to solve application problems (2)

1 In the question, the scores which all the players have gained added up should be an even number. It is absolutely impossible to be an odd number, so 1980 and 1984 are the two possible numbers. Let the number of contestants in the tournament be x. From the question, when the total score is 1980, 2 scores are awarded for each game, so there are 990 games, that is $\dfrac{x(x-1)}{2} = 990$, we get $x = 45$ (the negative value is not applicable, reject). When the total score is 1984, 2 scores are awarded for each game, so there are 992 games, that is $\dfrac{x(x-1)}{2} = 992$, the solution to the equation is a non-integer, reject it. Therefore, there are 45 contestants in the tournament. **2** (a) Let the width of the paths be x metres. From the question we get $18x + 15x - x^2 = 18 \times 15 \times \dfrac{1}{3}$, the solutions are $x_1 = 3$, $x_2 = 30$ (not applicable, reject). So the width of the paths in Diagram 1 is 3 metres. (b) Let the radius of the sector be y metres. From the question we get $\pi y^2 = 18 \times 15 \times \dfrac{1}{3}$, the solutions are $y_1 \approx 5.4$, $y_2 \approx -5.4$ (not applicable, reject). So the radius of the sectors in Diagram 2 is about 5.4 metres. **3** Let the two non-hypotenuse sides of the right-angled triangle be x m and y m $(x > y)$, then $\begin{cases} x - y = 1, \\ x^2 + y^2 = (x+1)^2 \end{cases}$, the solutions are $\begin{cases} x = 4, \\ y = 3 \end{cases}$. Therefore the hypotenuse of the triangle is 5 metres. **4** Let the original production plan each day be x tents and they are to be made in y days. From the question we get $\begin{cases} xy = 7200, \\ 2x(y-5) = 7200 \times 1.2 \end{cases}$ the solutions are $\begin{cases} x = 576, \\ y = 12.5 \end{cases}$. The factory produced

1152 tents each day. **5** Let to buy a metres of the original pure cotton fabric be x pounds and the percentage of the price of the pure cotton fabric increased be y. From the question we get $\begin{cases} x(1+y) = 36, \\ (60 - x)(1 - y) = 18 \end{cases}$, the solutions are $\begin{cases} x = 24, \\ y = \dfrac{1}{2} \end{cases}$, (the negative value is not applicable, reject). Therefore, the cost to buy a metres of the original pure cotton fabric was £24 and the percentage of the price of the pure cotton fabric was increased 50%.

Unit test 7

1 B   **2** C   **3** C   **4** D   **5** x^2, xy; $\dfrac{1}{2}$, -3; -10 **6** Answer may vary, for example: $xy = 2$, or $x^2 - xy = -1$

7 $\begin{cases} x_1 = 1, \\ y_1 = 2 \end{cases} \begin{cases} x_2 = 0, \\ y_2 = 3 \end{cases}$ **8** $(0, -2)$ or $(0, 6)$

9 $x - 3y = 0$, $x + y = 0$ **10** $\begin{cases} x = 3, \\ y = -1 \end{cases}$

11 $\begin{cases} x = -1, \\ y = -\dfrac{1}{2} \end{cases}$ and $\begin{cases} x = \dfrac{8}{5}, \\ y = \dfrac{4}{5} \end{cases}$ **12** $\begin{cases} x = 2 + \sqrt{3}, \\ y = -2\sqrt{3} \end{cases}$ and $\begin{cases} x = 2 - \sqrt{3}, \\ y = 2\sqrt{3} \end{cases}$ **13** Let the lengths of the two sides of the triangle be x metres and y metres $(x > y)$, then $\begin{cases} x^2 + y^2 = 10^2, \\ x + y = 14 \end{cases}$. The solutions are $\begin{cases} x = 6 \\ y = 8 \end{cases}$. Therefore the other two sides of the triangle are 6 metres and 8 metres. **14** Let the annual interest rate of Alvin's deposit be x and that of Wade's be y, $\begin{cases} 10\,000(1+x)^2 - 10\,000 + 209 = 10\,000(1+y)^2 - 10\,000, \\ 1.25x = y \end{cases}$, the solutions are $\begin{cases} x = 4\%, \\ y = 5\% \end{cases}$. The annual interest rate of Alvin's fixed savings account is 4% and that of Wade's is 5%.

Chapter 8   Vectors

8.1   Basic concepts of vectors

❶ B ❷ C ❸ B ❹ C ❺ B

❻ 2, –2 ❼ $\overrightarrow{OC}$ ❽ (a) $\overrightarrow{ED}$, $\overrightarrow{DC}$

(b) 6 ❾ (a) $\overrightarrow{AO}=\overrightarrow{OC}$, $\overrightarrow{BO}=\overrightarrow{OD}$

(b) $\overrightarrow{AB}=-\overrightarrow{CD}$, $\overrightarrow{DA}=-\overrightarrow{BC}$ ❿ (a) $\overrightarrow{DB}$, $\overrightarrow{FE}$

(b) $\overrightarrow{ED}$, $\overrightarrow{FA}$, $\overrightarrow{CF}$ (c) 7 ⓫ (a) $\overrightarrow{CO}$, $\overrightarrow{OF}$,

$\overrightarrow{DE}$ (b) $\overrightarrow{OA}$, $\overrightarrow{DO}$, $\overrightarrow{CB}$, $\overrightarrow{EF}$ (c) 9

8.2   Addition of vectors (1)

❶ B ❷ C ❸ C ❹ triangle law $\overrightarrow{AC}$ resultant ❺ zero vector **0** ❻ 3 km southwards, 2 km northwards, the original place

❼ (a) $\overrightarrow{AC}$ (b) $\overrightarrow{AO}$ (c) **0** (d) $\overrightarrow{AD}$

❽ (a) $\overrightarrow{AC}$, diagram correctly drawn (b) $\overrightarrow{ED}$, diagram correctly drawn (c) $\overrightarrow{EB}$, $\overrightarrow{CE}$ (d) **0**

❾ Diagrams correctly drawn ❿ D is at 9.

⓫ (a) $\dfrac{12}{5}$ (b) 5 (c) 5 (d) 5

8.3   Addition of vectors (2)

❶ C ❷ B ❸ (a) $\overrightarrow{AB}$ (b) $\overrightarrow{AE}$ ❹ $\sqrt{2}$

❺ **0** ❻ Diagrams correctly drawn (a) $\overrightarrow{AC}$

(b) $\overrightarrow{BE}$ (c) $\overrightarrow{FD}$ (d) $\overrightarrow{BE}$ (e) $\overrightarrow{BE}$ ❼ Diagram correctly drawn ❽ $2\sqrt{2}$ km northeastwards

❾ From the given we get $BO = \sqrt{3}CO$, so $\angle CBO = 30°$; then the sizes of the interior angles in rhombus $ABCD$ are 60°, 120°, 60°, 120°.

8.4   Subtraction of vectors (1)

❶ D ❷ B ❸ D ❹ B ❺ $\overrightarrow{BA}$ $\overrightarrow{AB}$

❻ $\overrightarrow{BC}$ $\overrightarrow{CD}$ ❼ (a) **b – a – c** (b) **b – a**

(c) **–a – c** ❽ $\sqrt{2}$ cm ❾ Diagrams correctly drawn ❿ (a) Correct (b) Wrong. The correct one should be: $\overrightarrow{CB}-\overrightarrow{AB}-\overrightarrow{CA}=\mathbf{0}$

⓫ (a) The original expression $=\overrightarrow{AB}+\overrightarrow{BD}+\overrightarrow{DC}+\overrightarrow{CA}=\mathbf{0}$ (b) The original expression $=\overrightarrow{BO}+\overrightarrow{OA}+\overrightarrow{AD}=\overrightarrow{BD}$ (c) The original expression $=\overrightarrow{NQ}+\overrightarrow{QP}+\overrightarrow{PM}+\overrightarrow{MN}=\mathbf{0}$ ⓬ (a) $\sqrt{5}$ (b) 2

(c) **0** (d) $\overrightarrow{AD}=2$

8.5   Subtraction of vectors (2)

❶ B ❷ B ❸ D ❹ **a + b** **b – c**

❺ 1 cm $2\sqrt{2}$ cm ❻ **–a**, **a + b**, **c – b**, **b – a**, **–a – b + c** ❼ $2\sqrt{2}$ ❽ Diagrams correctly drawn ❾ (a) $\overrightarrow{OC}$; $\overrightarrow{AO}$; $\overrightarrow{OB}$, $\overrightarrow{BO}$, $\overrightarrow{CA}$, $\overrightarrow{OD}$, $\overrightarrow{DO}$, $\overrightarrow{BD}$, $\overrightarrow{DB}$ (b) Diagram correctly drawn ❿ (a) $\overrightarrow{CE}$ $\overrightarrow{AC}$ (b) Diagrams correctly drawn of any four of $\overrightarrow{EB}$, $\overrightarrow{EC}$, $\overrightarrow{CE}$, $\overrightarrow{BC}$, $\overrightarrow{CB}$, $\overrightarrow{AD}$, $\overrightarrow{DA}$. (c) Diagram correctly drawn

⓫ (a) $\overrightarrow{BE}=2\overrightarrow{BC}$ (b) $\overrightarrow{DE}=2\overrightarrow{BC}-\dfrac{1}{2}\overrightarrow{BA}$

8.6   Multiplying a vector by a scalar (1)

❶ C ❷ C ❸ D ❹ **0** ❺ 6**a**

❻ $-\dfrac{1}{5}\mathbf{c}$ ❼ $m\mathbf{a}+n\mathbf{b}$ ❽ 6; $\overrightarrow{AB}$, $\overrightarrow{BA}$, $\overrightarrow{AC}$, $\overrightarrow{CA}$, $\overrightarrow{BC}$, $\overrightarrow{CB}$ ❾ $m = 0$ or $n = 0$ or $\mathbf{a}=\mathbf{0}$

❿ Take a point O in the plane, then construct $\overrightarrow{OA} = \mathbf{a}$. Take $OB = \dfrac{4}{3}OA$ on the ray OA, then $\overrightarrow{OB}=\dfrac{4}{3}\mathbf{a}$. ⓫ Through point B construct a line parallel to AD, intersecting EF at point G, and intersecting DC at point H. It is easy to prove quadrilateral $ABHD$ is a parallelogram. Since $EF /\!/ DC$, $AB = FG = DH$. Since $\overrightarrow{AB}$, $\overrightarrow{FE}$ and $\overrightarrow{DC}$ are the vectors with the same direction, and since $5AB = 3CD$, $\overrightarrow{FG} = \overrightarrow{AB} = \dfrac{3}{5}\overrightarrow{DC}$. In $\triangle BCH$, $GE /\!/ HC$, $\dfrac{GE}{HC} = \dfrac{BE}{BC}$ and $BE : EC = 1 : 2$, $BE : BC = 1 : 3$, therefore $GE = \dfrac{1}{3}HC$, $HC = \dfrac{2}{5}DC$, $GE = \dfrac{2}{15}DC$.

$\overrightarrow{GE} = \dfrac{2}{15}\overrightarrow{DC}$, $\overrightarrow{FE} = \overrightarrow{FG}+\overrightarrow{GE} = \dfrac{3}{5}\overrightarrow{DC}+\dfrac{2}{15}\overrightarrow{DC} = \dfrac{11}{15}\overrightarrow{DC}$ ⓬ Take a point O in the plane, then construct ray OM with the same direction as **b**. Take $OC = \sqrt{5}\,|\,\mathbf{b}\,|$, then $\overrightarrow{OC} = \sqrt{5}\mathbf{b}$. Construct ray CN in the opposite direction of **a**. Take $CD =$

$\dfrac{2}{3}$ | **a** | , then $\overrightarrow{CD} = -\dfrac{2}{3}$ **a**. Connect OD. Vector $\overrightarrow{OD}$ is $\sqrt{5}$ **b** $- \dfrac{2}{3}$ **a**, which is the required construction.

8.7 Multiplying a vector by a scalar (2)

① D　**②** C　**③** B　**④** D　**⑤** 60**a** associative　**⑥** 5**a** $+ 5$**b** distributive　**⑦** 2**a** $+$ **b** **⑧** $(m-n)$**a**　**⑨** -14**a** $+ \dfrac{17}{2}$**b**　**⑩** 3**a** $+ 2$**b** $+ 6$**c**　**⑪** **x** $= \dfrac{1}{2}$**a** $- \dfrac{11}{6}$**b**　**⑫** Construct $\overrightarrow{OB} = \dfrac{2}{3}$**a**, $\overrightarrow{BC} = -$**b**, then $\overrightarrow{OC} = \dfrac{1}{3}(2$**a** $- 3$**b**$)$

⑬ $(m-n)(2$**a** $- 3$**x**$) = 4$**b**, $(2m-2n)$**a** $- 3(m-n)$**x** $= 4$**b**, $3(m-n)$**x** $= (2m-2n)$**a** $- 4$**b**. Since $m \neq n$, **x** $= \dfrac{2}{3}$**a** $- \dfrac{4}{3(m-n)}$**b**.

8.8 Multiplying a vector by a scalar (3)

① C　**②** C　**③** A　**④** A　**⑤** is parallel to　**⑥** $AB \mathbin{/\mkern-3mu/} CD$ or on the same line, $AB = \dfrac{3}{2}CD$　**⑦** ± 3　**⑧** $-\dfrac{1}{k}$**a**　**⑨** 8　**⑩** in opposite directions　**⑪** Since **a** $+ 2$**b** $= 3$**c**, **a** $= -2$**b** $+ 3$**c**; since **b** $+ \dfrac{20}{3}$**c** $= 2$**a**, **a** $= \dfrac{1}{2}$**b** $+ \dfrac{10}{3}$**c**, -2**b** $+ 3$**c** $= \dfrac{1}{2}$**b** $+ \dfrac{10}{3}$**c** and **b** $= -\dfrac{2}{15}$**c**, that is, **b** $\mathbin{/\mkern-3mu/}$ **c**, and they are in opposite directions.　**⑫** From $AD \mathbin{/\mkern-3mu/} EF \mathbin{/\mkern-3mu/} BC$, we get $\overrightarrow{AD} \mathbin{/\mkern-3mu/} \overrightarrow{EF} \mathbin{/\mkern-3mu/} \overrightarrow{BC}$ and $EF = \dfrac{AD + BC}{2} = \dfrac{9}{2}$. $\overrightarrow{CB}$ is in the opposition direction of **a**, $\overrightarrow{EF}$ is in the same direction as **a**, | **a** | $= 3$, | $\overrightarrow{CB}$ | $= 6$, | $\overrightarrow{EF}$ | $= \dfrac{9}{2}$. | $\overrightarrow{CB}$ | $= 2$| **a** |, | $\overrightarrow{EF}$ | $= \dfrac{3}{2}$| **a** |, therefore $\overrightarrow{CB} = -2$**a** and $\overrightarrow{EF} = \dfrac{3}{2}$**a**. Through point D construct $DG \perp CB$ at point G. It is easy to prove that quadrilateral $ABGD$ is a rectangle, $AB = DG$ and $BG = GC = AD$. $\overrightarrow{GC}$ is in the same direction as **a**, then $\overrightarrow{GC} =$ **a**. Since $\overrightarrow{AB} =$ **b**, and $\overrightarrow{DG}$ and $\overrightarrow{AB}$ are in the same direction, $\overrightarrow{DG} =$

b, so $\overrightarrow{DC} =$ **a** $+$ **b**.　**⑬** (a) $\overrightarrow{AB} = 4$**a**, $\overrightarrow{DC} = 4$**a**, $\overrightarrow{BC} = 2$**b**　(b) 8**a** $+ 4$**b**　(c) $\overrightarrow{HE} = 2$**a** $-$ **b**, $\overrightarrow{EF} = 2$**a** $+$ **b**.

8.9 Linear operations on vectors (1)

① D　**②** A　**③** C　**④** D　**⑤** $\dfrac{2}{3}$**a** $-$ **b** **⑥** $\dfrac{1}{4}$**a** $+ \dfrac{3}{4}$**b**　**⑦** $\dfrac{2}{3}$**b** $- \dfrac{2}{3}$**a**　**⑧** $\dfrac{1}{2}$**a** $+ \dfrac{1}{2}$**b** **⑨** 3**b** $- 3$**a**　**⑩** 2**a** $- 2$**b**　**⑪** Since $\overrightarrow{EF} = \overrightarrow{EA} + \overrightarrow{AB} + \overrightarrow{BF}$ and $\overrightarrow{EF} = \overrightarrow{ED} + \overrightarrow{DC} + \overrightarrow{CF}$ and $\overrightarrow{EA} = -\overrightarrow{ED}$ and $\overrightarrow{BF} = -\overrightarrow{CF}$, $2\overrightarrow{EF} = \overrightarrow{AB} + \overrightarrow{DC}$.　**⑫** (a) $\overrightarrow{AD} = -6$**a**　(b) Since $\overrightarrow{BC} = -4$**a** and $\overrightarrow{AD} = -6$**a**, we get $\overrightarrow{BC} \mathbin{/\mkern-3mu/} \overrightarrow{AD}$, i. e., sides AD and BC are parallel; since $\overrightarrow{AB} =$ **a** $+ 2$**b**, and $\overrightarrow{CD} = -3$**a** $- 2$**b** (**a** and **b** are not parallel), $\overrightarrow{AB}$ is not parallel to $\overrightarrow{CD}$, i. e., sides AB and CD are not parallel. Therefore, quadrilateral $ABCD$ is a trapezium.

8.10 Linear operations on vectors (2)

① C　**②** B　**③** B　**④** B　**⑤** $\dfrac{4}{7}$**a** $- \dfrac{4}{7}$**b** **⑥** $\dfrac{1}{2}$**a** $- \dfrac{1}{2}$**b**, $\dfrac{1}{2}$**a** $+ \dfrac{1}{2}$**b**　**⑦** 2**b** $-$ **a**　**⑧** $\dfrac{1}{2}$**a** $+ \dfrac{1}{2}$**b**　**⑨** $\overrightarrow{FD} = \dfrac{4}{5}$**a** $+ \dfrac{1}{5}$**b**, $\overrightarrow{FC} = \dfrac{1}{5}$**b** $- \dfrac{1}{5}$**a** **⑩** Since $ABCD$ is a parallelogram, $AB \mathbin{/\mkern-3mu/} CD$, $AD \mathbin{/\mkern-3mu/} BC$, $AB = CD$ and $AD = BC$. Since $\overrightarrow{BA} =$ **a** and $\overrightarrow{BC} =$ **b**, $\overrightarrow{CD} =$ **a** and $\overrightarrow{AD} =$ **b**. And since $AE = 3ED$, $\overrightarrow{ED} = \dfrac{1}{4}$**b**, $\overrightarrow{AE} = \dfrac{3}{4}$**b** and $\overrightarrow{CE} = \overrightarrow{CD} + \overrightarrow{DE} =$ **a** $- \dfrac{1}{4}$**b**. Since $EF = CE$, $\overrightarrow{EF} = \overrightarrow{CE} =$ **a** $- \dfrac{1}{4}$**b**, therefore $\overrightarrow{AF} = \overrightarrow{AE} + \overrightarrow{EF} = \dfrac{3}{4}$**b** $+$ **a** $- \dfrac{1}{4}$**b** $=$ **a** $+ \dfrac{1}{2}$**b**.

8.11 Column vectors and their operations (1)

① B　**②** D　**③** D　**④** $\begin{pmatrix} 3 \\ -5 \end{pmatrix}$　**⑤** 5 [Hint: 2**a** $-$ **b** $= \begin{pmatrix} 3 \\ -4 \end{pmatrix}$]　**⑥** $\begin{pmatrix} 4 \\ -2 \end{pmatrix}$　**⑦** $\sqrt{10}$

$\left[\text{Hint:} \ \overrightarrow{AC} = \overrightarrow{AB} + \overrightarrow{BC} = \begin{pmatrix} 3 \\ -1 \end{pmatrix}\right]$ **8** $\begin{pmatrix} -1 \\ -2 \end{pmatrix}$

$\left[\text{Hint:} \ \overrightarrow{CD} = \overrightarrow{BA} = \overrightarrow{OA} - \overrightarrow{OB} = \begin{pmatrix} 2 \\ -1 \end{pmatrix} - \begin{pmatrix} 3 \\ 1 \end{pmatrix}\right]$

9 $\begin{pmatrix} 5 \\ 0 \end{pmatrix}$ [Hint: Let C be $(x, 0)$, then $\overrightarrow{AB} \parallel \overrightarrow{AC}$,

that is $\begin{pmatrix} 2 \\ -1 \end{pmatrix} /\!/ \begin{pmatrix} x-1 \\ -2 \end{pmatrix}$, so $\dfrac{2}{x-1} = \dfrac{-1}{-2}$, and $x = 5$]

10 (a) $\begin{pmatrix} -2 \\ 16 \end{pmatrix}$ (b) $\sqrt{65}$ [Hint: (a) $2\mathbf{a} +$

$3\mathbf{b} - 3\mathbf{c} = \begin{pmatrix} 2 \times 2 + 3 \times 1 - 3 \times 3 \\ 2 \times (-1) + 3 \times 2 - 3 \times (-4) \end{pmatrix} = \begin{pmatrix} -2 \\ 16 \end{pmatrix}$

(b) $3\mathbf{a} - 2\mathbf{b} = \begin{pmatrix} 4 \\ -7 \end{pmatrix}$, so $|3\mathbf{a} - 2\mathbf{b}| = \sqrt{65}$]

11 (a) $\mathbf{a} = \begin{pmatrix} 1 \\ 1 \end{pmatrix}$, $\mathbf{b} = \begin{pmatrix} 1 \\ -1 \end{pmatrix}$ [Hint: (a) $3\mathbf{b} =$

$\mathbf{a} + 2\mathbf{b} - (\mathbf{a} - \mathbf{b}) = \begin{pmatrix} 3 \\ -3 \end{pmatrix}$, hence $\mathbf{b} = \begin{pmatrix} 1 \\ -1 \end{pmatrix}$, and

$\mathbf{a} = \mathbf{b} + \begin{pmatrix} 0 \\ 2 \end{pmatrix} = \begin{pmatrix} 1 \\ 1 \end{pmatrix}$]   (b) $3\mathbf{a} - 2\mathbf{c} = \begin{pmatrix} 3 \\ 3 \end{pmatrix} -$

$\begin{pmatrix} 6 \\ -8 \end{pmatrix} = \begin{pmatrix} -3 \\ 11 \end{pmatrix}$, therefore $|3\mathbf{a} - 2\mathbf{c}| = \sqrt{130}$

12 (a) $(4, 0)$ (b) $\sqrt{65}$ (c) $\begin{pmatrix} -\dfrac{3}{2} \\ 0 \end{pmatrix}$ [Hint:

(a) $\overrightarrow{AD} = \overrightarrow{BC} = \begin{pmatrix} 2 \\ -2 \end{pmatrix}$, $\overrightarrow{OD} = \overrightarrow{OA} + \overrightarrow{AD} = \begin{pmatrix} 4 \\ 0 \end{pmatrix}$,

hence $D(4, 0)$;   (b) $\overrightarrow{BD} = \begin{pmatrix} 4 \\ 0 \end{pmatrix} - \begin{pmatrix} -3 \\ 4 \end{pmatrix} =$

$\begin{pmatrix} 7 \\ -4 \end{pmatrix}$, hence $|BD| = \sqrt{65}$;   (c) $\overrightarrow{OG} = \dfrac{1}{2}(\overrightarrow{OA} +$

$\overrightarrow{OC}) = \begin{pmatrix} \dfrac{1}{2} \\ 2 \end{pmatrix}$, $\overrightarrow{AG} = \overrightarrow{OG} - \overrightarrow{OA} = \begin{pmatrix} \dfrac{1}{2} \\ 2 \end{pmatrix} - \begin{pmatrix} 2 \\ 2 \end{pmatrix} =$

$\begin{pmatrix} -\dfrac{3}{2} \\ 0 \end{pmatrix}$]

8.12 Column vectors and their operations (2)

1 B   **2** D   **3** C   **4** D   **5** D

6 $\sqrt{x_1^2 + y_1^2}$, $\sqrt{x_2^2 + y_2^2}$, $\sqrt{(x_1 - x_2)^2 + (y_1 - y_2)^2}$

7 $\mathbf{a} /\!/ \mathbf{b} /\!/ \mathbf{d}$ **8** $\pm \begin{pmatrix} \sqrt{5} \\ 2\sqrt{5} \end{pmatrix}$ **9** -1 or $\dfrac{3}{2}$

10 $\dfrac{1}{2}$ **11** (a) $(0, 2)$ [Hint: Let D be (x, y),

from $\overrightarrow{DC} = \overrightarrow{AB}$ we get $\begin{pmatrix} 1 - x \\ 3 - y \end{pmatrix} = \begin{pmatrix} 1 \\ 1 \end{pmatrix}$; therefore

$x = 0$, and $y = 2$]   (b) the column vector is

$\begin{pmatrix} -1 \\ -1 \end{pmatrix}$, since $\overrightarrow{CB} + \begin{pmatrix} -1 \\ -1 \end{pmatrix} = \overrightarrow{DA}$ (c) 2

12 (1) $\begin{pmatrix} 3 \\ 4 \end{pmatrix}$, $\begin{pmatrix} -1 \\ 0 \end{pmatrix}$ or $\begin{pmatrix} 1 \\ 0 \end{pmatrix}$ (2) 4

13 (a) Establish a coordinate plane as shown.

We can get the coordinates of E and F: $E\left(\dfrac{d}{2}, \dfrac{y}{2}\right)$

and $F\left(\dfrac{b+c}{2}, \dfrac{y}{2}\right)$. Using column vectors, we can

get $\overrightarrow{EF} = \begin{pmatrix} \dfrac{b+c-d}{2} \\ 0 \end{pmatrix}$, $\overrightarrow{AB} = \begin{pmatrix} b \\ 0 \end{pmatrix}$, and $\overrightarrow{DC} =$

$\begin{pmatrix} c - d \\ 0 \end{pmatrix}$, therefore $EF /\!/ AB /\!/ DC$.

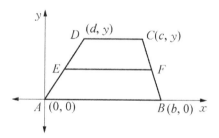

(b) From (a), we get $\overrightarrow{EF} = \begin{pmatrix} \dfrac{b+c-d}{2} \\ 0 \end{pmatrix}$, and

$\overrightarrow{AB} + \overrightarrow{DC} = \begin{pmatrix} b \\ 0 \end{pmatrix} + \begin{pmatrix} c - d \\ 0 \end{pmatrix} = \begin{pmatrix} b+c-d \\ 0 \end{pmatrix}$, so

$EF = \dfrac{1}{2}(AB + DC)$.

Unit test 8

1 B   **2** D   **3** B   **4** C   **5** $\overrightarrow{AD} - \overrightarrow{AB}$

$\overrightarrow{AC} - \overrightarrow{AD}$ **6** $\mathbf{b} - \dfrac{3}{2}\mathbf{a}$ **7** $\begin{pmatrix} 6 \\ 6 \end{pmatrix}$, 5   **8** (18,

14)   **9** $\begin{pmatrix} 3 \\ 1 \end{pmatrix}$ **10** $\begin{pmatrix} -\dfrac{7}{3} \\ -8 \end{pmatrix}$ **11** Diagrams

correctly drawn   **12** (a) $\overrightarrow{OC} = -\mathbf{a}$ (b) $\overrightarrow{OD} =$

$-\mathbf{b}$ (c) $\overrightarrow{AB} = \mathbf{b} - \mathbf{a}$ (d) $\overrightarrow{BC} = -\mathbf{a} - \mathbf{b}$

13 $\pm(\sqrt{2}, \sqrt{2})$ **14** (a) points correctly drawn;

quadrilateral *ABCD* correctly drawn　(b) $\begin{pmatrix} 5 \\ 5 \end{pmatrix}$

(c) Using column representation of the vectors, we can get $\overrightarrow{AB} = \begin{pmatrix} 2 \\ 5 \end{pmatrix}$, $\overrightarrow{BC} = \begin{pmatrix} 3 \\ 0 \end{pmatrix}$, $\overrightarrow{CD} = \begin{pmatrix} -2 \\ -5 \end{pmatrix}$ and $\overrightarrow{AD} = \begin{pmatrix} 3 \\ 0 \end{pmatrix}$, therefore $\overrightarrow{AB} /\!/ \overrightarrow{CD}$ and $\overrightarrow{BC} /\!/ \overrightarrow{AD}$. Hence quadrilateral *ABCD* is a parallelogram.

Chapter 9　Circles and properties of circles

9.1　Arc length

① D　② C　③ A　④ 72°　⑤ 6.28　⑥ 4　⑦ 3　⑧ 13.14　⑨ 9.42 cm　8 cm　30°　⑩ Ming walked 31.4 metres, and Larry also walked 31.4 metres.　⑪ 60 cm

⑫ 33.12 cm [Hint: the perimeter of the shaded part = $\frac{1}{4}$ × the circumference of the bigger circle + $\frac{1}{4}$ × the circumference of the smaller circle + $2(DC - AD)$]　⑬ 39.4 cm [Hint: the arc length of the bigger circle is 6π cm, and the arc length of the smaller circle is 4π cm]

9.2　The area of a sector

① C　② B　③ D　④ 270°　339.12 mm² 2.512 cm　2.512 cm²　5 cm　5.23 cm　10 m. 5π m²　⑤ 37.68　⑥ 3.768　⑦ 720 ⑧ 25　⑨ 53.24 cm²　⑩ 10.26 cm²

⑪ 11.44 cm²　⑫ 171.5 cm² [Hint: the area of the shaded part = the area of △*ACD* – the small area of the sector, the area of the small sector = $\frac{1}{4} \times 3.14 \times 10^2 - \frac{1}{2} \times 10^2 = 28.5$ (cm²)]

⑬ 57 cm² [Hint: the area of the shaded part = the area of sector *ABD* – the area of △*ABC*]

9.3　Determining a circle

① A　② B　③ C　④ (a) >　(b) = (c) <　⑤ (a) inside　(b) on　(c) outside ⑥ infinitely many, perpendicular bisector ⑦ one, perpendicular bisectors　⑧ $(x - a)^2 +$

$(y - b)^2 = r^2$　⑨ $OP = \sqrt{65} > 8$, hence point *P* is outside the circle.　⑩ Let the coordinates of the intersecting point be $(x, x + 1)$. Then we can get the equation $(x - 2)^2 + (x + 1 - 3)^2 = 18$; the solutions are $x_1 = 5$, $x_2 = -1$; therefore the coordinates of the intersecting point are $(5, 6)$ and $(-1, 0)$. Line and circle correctly drawn.

⑪ From the given we can get $BF = 4\sqrt{3}$ and $BE = \frac{7}{2}\sqrt{3}$. Join *AP*. When $BP = AP$, point *A* is on the circle with centre *P* and △*ABP* ∽ △*ABF*, so $BP = AP = \frac{4}{3}\sqrt{3}$. Therefore when $BP > AP$, that is $BP > \frac{4}{3}\sqrt{3}$, point *A* is inside the circle with centre *P*. And since point *E* is outside the circle, $BP < \frac{1}{2}BE$, that is, $BP < \frac{7}{4}\sqrt{3}$. Therefore, when $\frac{4}{3}\sqrt{3} < BP < \frac{7}{4}\sqrt{3}$, point *A* is inside the circle with centre *P* and point *E* is outside the circle.

9.4　Central angle, arc, chord, distance from chord to centre and their relationships (1)

① B　② B　③ B　④ B　⑤ an arc, a chord　⑥ 60　$\overset{\frown}{BC}$ and $\overset{\frown}{CD}$ perpendicular ⑦ twice　⑧ 110°　⑨ 15°　⑩ Join *AO* and *BO*. Since △*AOC* ≅ △*BOD*, ∠*AOE* = ∠*BOF*; therefore $\overset{\frown}{AE} = \overset{\frown}{BF}$.　⑪ Join *OC*. Since right-angled △*COD* ≅ right-angled △*COE*, ∠*AOC* = ∠*BOC*; therefore $\overset{\frown}{AC} = \overset{\frown}{BC}$.　⑫ Join *OE*. △*OED* is a right – angled △, and $OD = \frac{1}{2}OE$, that is ∠*OED* = 30°, so ∠*COE* = 2∠*EOA*, that is, $\overset{\frown}{CE} = 2\overset{\frown}{AE}$.

9.5　Central angle, arc, chord, distance from chord to centre and their relationships (2)

① D　② B　③ (a) ∠*AOB* = ∠*COD*　*OE* = *OF*　$\overset{\frown}{AB} = \overset{\frown}{CD}$　(b) *AB* = *CD*　∠*AOB* = ∠*COD*　$\overset{\frown}{AB} = \overset{\frown}{CD}$　(c) *AB* = *CD*　∠*AOB* =

$\angle COD$ $OE = OF$ (d) $AB = CD$ $\overset{\frown}{AB} =$ $\overset{\frown}{CD}$ $OE = OF$ ❹ 3   ❺ 40° 80°   ❻ 65°

❼ $\dfrac{4\pi}{3}$ ❽ $\dfrac{2}{3}\pi - \sqrt{3}$ ❾ Join OC, so $\triangle CDO \cong$ $\triangle CEO(\text{SAS})$, $CD = CE$.   ❿ $AD = BC$, $\overset{\frown}{AD} =$ $\overset{\frown}{BC}$, that is, $\overset{\frown}{AB} = \overset{\frown}{DC}$, $AB = DC$ ⓫ It is easy to prove $\angle BOC = 30°$, $OB \perp AC$ and $BD \perp CO$. Therefore $\angle AKD = 150°$.

9.6　Perpendicular chord bisector theorem (1)

❶ A   ❷ D   ❸ B   ❹ 5   ❺ 4   ❻ 3 or 27   ❼ 6 or $2\sqrt{21}$ ❽ Through point O draw $OE \perp AB$, then $AE = BE$. Since $CE = DE$, $AC = BD$.   ❾ 10   ❿ Through point O draw $OH \perp EF$ at H. Join OE and OF. From $OE = 2$ and $EH = \dfrac{1}{2}EF = \sqrt{3}$, we can get $OH = 1$. Since $\sin \angle ABC = \dfrac{OH}{BO} = \dfrac{1}{3}$, we have $OB = 3$.

⓫ Join AB. Let $DO = k$, then $AO = 5 - k$, $BO = 2k$. In the right-angled $\triangle ABO$, $AO^2 + BO^2 = AB^2$, that is, $(5 - k)^2 + (2k)^2 = 25$, therefore $k = 2$, and $D(0, 2)$

9.7　Perpendicular chord bisector theorem (2)

❶ C   ❷ A   ❸ B   ❹ B   ❺ (a) $\angle AOD$, $\angle BOD$, $\perp$, $=$ (b) $=$, $\angle AOD$, $\angle BOD$, $\perp$ ❻ $2\sqrt{3}$ ❼ $\sqrt{10}$ or $3\sqrt{10}$ ❽ 4 and 6   ❾ 30° or 90°   ❿ Diagram correctly drawn   ⓫ (a) Draw $DH \perp CE$ with point H being the perpendicular foot. Since D is the centre of the semicircle, $AC = 5$ and $AE = 1$, $CH = \dfrac{1}{2}EC = 2$. Since $AB = AC$, $\angle B = \angle C$, thus $\cos C = \cos B = \dfrac{4}{5}$. In right-angled $\triangle CDH$, since $\cos C = \dfrac{CH}{CD} = \dfrac{4}{5}$ and $CH = 2$, $CD = \dfrac{5}{2}$ (b) Draw $AM \perp BC$ with point M being the perpendicular foot. Join AF. Since $CD = \dfrac{5}{2}$, $CF = 5$. In right-angled

$\triangle ACM$, since $\cos C = \dfrac{CM}{AC} = \dfrac{4}{5}$ and $AC = 5$, $CM = $ 4, therefore $AM = \sqrt{AC^2 - CM^2} = \sqrt{5^2 - 4^2} = $ 3. From $CF = 5$ and $CM = 4$, $FM = 1$, therefore $AF = \sqrt{AM^2 + FM^2} = \sqrt{3^2 + 1^2} = \sqrt{10}$.

⓬ (a) 16 m   (b) 2 m

9.8　The relationship between a line and a circle

❶ C   ❷ C   ❸ C   ❹ C   ❺ ① 10   ② 2   ③ no   ❻ one   ❼ $3\sqrt{3}$ ❽ 4 cm   ❾ $5 < r \leqslant 12$ or $r = \dfrac{60}{13}$ ❿ Through point A draw $AD \perp BC$. Since $AB = AC$, $\angle BAD = 60°$, $BD = 2\sqrt{3}$ and $AD = 2$, therefore BC and the circle are tangent to each other.   ⓫ (a) From $OP \perp AB$, we get $\angle OPA = \angle OPB = 90°$. Hence $\angle POB = 90° - \angle OBA$. Since $\angle PAO = 90° - \angle OBA$, we get $\angle POB = \angle PAO$. Therefore, $\triangle OBP$ is similar to $\triangle AOP$.   (b) $(\sqrt{2}, \sqrt{2})$ ⓬ $(2, 5)$ or $(2, 1)$

9.9　The relationship between two circles (1)

❶ C   ❷ A   ❸ A   ❹ 2   1   zero   ❺ 9 cm or 1 cm   ❻ $1 + 2\sqrt{2}$ ❼ $3 < t < 5$ or $7 < t < 9$ ❽ Let the radii of the three circles be x, y and z. Then we have $\begin{cases} x + y = 6 \\ y + z = 8 \\ z + x = 10, \end{cases}$ therefore the radii of the three circles are 4, 2 and 6.   ❾ Join OB. In the right-angled $\triangle OCB$, $OC = 6$ and $BC = 8$, so $OB = 10$. The sum of the radii of the two circles is 10, hence they are tangent to each other.   ❿ Since point P is inside the circle with centre O, the two circles, centres O and P, can only be tangent to each other internally. Since O is the midpoint of AB and P is the midpoint of BC, $OP = \dfrac{1}{2}AC = 3$ cm. Since the circle, centre O, is the circumscribed circle of $\triangle ABC$, its radius is 5 cm; the radius of the circle with centre P is PQ. When the circle, centre P, is tangent to the circle, centre O, we have $|5 - PQ| = 3$. When $5 - PQ = 3$, $PQ = 2$ cm, $t = 1$ second; when $5 - PQ = -3$, $PQ = 8$ cm, $t = 4$ second. Hence the value of t is either 1 second or 4 seconds.　⑪ $t = 2$ seconds

9.10　The relationship between two circles (2)

① C　② C　③ A　④ B　⑤ 3 or 13

⑥ $\dfrac{8}{5} < r < \dfrac{32}{5}$　⑦ $\dfrac{3}{5}$　⑧ $3 - \sqrt{5} < x \leqslant 3 + \sqrt{5}$　⑨ 9 or 21　⑩ Let the radius of the circle, centre B, be R. Then $5 + R = 12$ or $|R - 5| = 12$; therefore $R = 7$ or $R = 17$　⑪ 1 or 7

⑫ $\left(\dfrac{22}{5}, \dfrac{24}{5}\right)$ or $\left(-\dfrac{18}{5}, -\dfrac{6}{5}\right)$ or $\left(-\dfrac{42}{5}, -\dfrac{24}{5}\right)$ or $\left(-\dfrac{2}{5}, \dfrac{6}{5}\right)$

Unit test 9

① B　② C　③ C　④ C　⑤ A　⑥ B

⑦ B　⑧ 22.28　⑨ 5.024　⑩ 84.78

⑪ 64　⑫ the diameter　⑬ half　⑭ 8

⑮ equal　⑯ 50° 40°　⑰ $6 + 4\sqrt{3}$

⑱ $\dfrac{3}{2}$　⑲ Since M and N are the midpoints of the chords AB and CD, we can get $OM \perp AB$, and $ON \perp CD$. Since $\angle OMN = \angle ONM$, we can get $OM = ON$. Therefore $AB = CD$.　⑳ (a) 30°
(b) From $\overset{\frown}{CD} = \overset{\frown}{DE} = \overset{\frown}{EF} = \overset{\frown}{FB}$, we can get $\overset{\frown}{CE} = \overset{\frown}{EB}$, so $\angle COE = \angle BOE$, i.e., OE bisects $\angle COB$. So OE also bisects BC. Therefore, $OE \perp CB$.　㉑ (a) 80 cm　(b) 45°　㉒ 2.43 cm^2
㉓ (a) In rectangle $ABCD$, since $AD \parallel BC$, $\angle APB = \angle DAP$. From the question we get $\angle QAD = \angle DAP$, then $\angle APB = \angle QAD$. Since $\angle B = \angle ADQ = 90°$, $\triangle ADQ \backsim \triangle PBA$, so $\dfrac{DQ}{AB} = \dfrac{AD}{BP}$, that is, $\dfrac{y}{3} = \dfrac{4}{x + 4}$, therefore $y = \dfrac{12}{x + 4}$, and the domain is $x > 0$.　(b) No, it won't. Since $\angle QAD = \angle DAP$, $\angle ADE = \angle ADQ = 90°$ and $AD = AD$, therefore $\triangle ADE \cong \triangle ADQ$. Hence

$DE = DQ = y$. $S = S_{\triangle AQE} + S_{\triangle PQE} = \dfrac{1}{2}QE \times AD + \dfrac{1}{2}QE \times PC = \dfrac{48}{x + 4} + \dfrac{12x}{x + 4} = 12$.　(c) Through point Q draw $QF \perp AP$ at point F. Since the circle, centre Q, with radius 4 is tangent to line AP, $QF = 4$. Since $S = 12$, $AP = 6$. In the right-angled $\triangle ABP$, since $AB = 3$, $\angle BPA = 30°$ and $\angle PAQ = 60°$, so $AQ = \dfrac{8\sqrt{3}}{3}$. Let the radius of the circle, centre A, be r.　(i) When the circle, centre A, is tangent to the circle, centre Q, externally, $AQ = r + 4$, that is, $\dfrac{8\sqrt{3}}{3} = r + 4$, so $r = \dfrac{8\sqrt{3}}{3} - 4$.　(ii) When the circle, centre A, is tangent to the circle, centre Q, internally, $AQ = r - 4$, that is, $\dfrac{8\sqrt{3}}{3} = r - 4$, so $r = \dfrac{8\sqrt{3}}{3} + 4$. In summary, the radius of the circle, centre A, is $\dfrac{8\sqrt{3}}{3} - 4$ or $\dfrac{8\sqrt{3}}{3} + 4$.

End of year test

① A　② B　③ D　④ B　⑤ D　⑥ D

⑦ B　⑧ D　⑨ $\dfrac{3}{2}\pi$　⑩ infinitely many

$\begin{cases} x = 1, \\ y = 3, \end{cases} \begin{cases} x = 3, \\ y = 2, \end{cases}$ and $\begin{cases} x = 5, \\ y = 1 \end{cases}$　⑪ -3　9

⑫ 3　⑬ $\dfrac{y^2}{x^6 z^4}$　⑭ $\dfrac{1}{2a}$　⑮ $x_1 = 0$, $x_2 = 3$

⑯ zero　⑰ 0　⑱ $\begin{pmatrix} 2 \\ -2 \end{pmatrix}$　$2\sqrt{2}$　⑲ 70°
35°　⑳ $3(2a - b)(a + b)(a - 5b)$　㉑ $x = -\dfrac{1}{2}$　㉒ $-\dfrac{1}{x + 1}$, $-\dfrac{3}{7}$　㉓ $(7^{\frac{3}{2}} \times 49^{-\frac{3}{4}})^{\frac{1}{3}} = (7^{\frac{3}{2}} \times 7^{-\frac{3}{2}})^{\frac{1}{3}} = (7^0)^{\frac{1}{3}} = 1$　㉔ The original expression $= (9\sqrt{2} + \sqrt{2} - 2\sqrt{2}) \div 4\sqrt{2} = 8\sqrt{2} \div 4\sqrt{2} = 2$　㉕ $\angle \theta = 18°$　㉖ $x_1 = \dfrac{4}{3} + \dfrac{\sqrt{7}}{3}$, $x_2 = \dfrac{4}{3} - \dfrac{\sqrt{7}}{3}$　㉗ $\begin{cases} x = 5 \\ y = 10 \end{cases}$ and $\begin{cases} x = -5 \\ y = -10 \end{cases}$

28 (a) Substituting $x = -1$ into the equation, we get $1 + m - 2 = 0$, so $m = 1$. The equation is $x^2 - x - 2 = 0$ and the solution is $x_1 = 2$ and $x_2 = -1$. Therefore, the other root of the equation is $x = 2$ (b) From the given we get $\Delta = m^2 + 8$ and $m^2 \geqslant 0$, so $m^2 + 8 > 0$, that is, $\Delta > 0$. Therefore the equation has two unequal real roots. **29** (a) $\mathbf{a} - \mathbf{b}$, $-\mathbf{a} - \mathbf{b}$ (b) Diagram correctly drawn

30 9.28 cm **31** 200 km, $\dfrac{9}{2}$ hours **32** Let the price increase be x pounds. The original profit from selling each unit was $120 - 100 = £20$. From the given we have the equation $(x + 20)(500 - 10x) = 12\,000$; the solutions are $x_1 = 10$ and $x_2 = 20$. When $x_1 = 10$, $120 + 10 = 130$, $500 - 100 = 400$; when $x_2 = 20$, $120 + 20 = 140$, $500 - 200 = 300$. Answer: when the price was set at £130 per unit, the shop should buy 400 units; when the price was set at £140 per unit, the shop should buy 300 units. The shop would earn a total profit of £12\,000 from either price. **33** (a) Draw $OH \perp AP$ at point H. Since OH passes through the centre, AP is the chord, $AP = 2AH$. In right-angled $\triangle AOH$, since $\tan A = \dfrac{1}{2}$ and $OA = 3$, let $OH = k$ and $AH = 2k$. From $AO^2 = OH^2 + AH^2$, we get $k = \dfrac{3}{5}\sqrt{5}$, then $AP = 2AH = \dfrac{12}{5}\sqrt{5}$.

(b) Join PO and join OQ. Since the circle, centre Q, passes through points P and O, $PQ = OQ$, and $\angle QPO = \angle QOP$. Since a circle, centre O, passes through points P and A, $PO = AO$, and $\angle QPO = \angle A$, so $\angle QOP = \angle A$. Since $\angle P = \angle P$, $\triangle QPO \backsim \triangle OPA$, and $\dfrac{AP}{OP} = \dfrac{AO}{QO}$, that is, $\dfrac{x}{3} = \dfrac{3}{y}$, therefore $y = \dfrac{9}{x}$, $0 < x \leqslant 6$.